The Enemy of Good

Estimating the Cost of Waiting for Nearly Perfect Automated Vehicles

Nidhi Kalra and David G. Groves

For more information on this publication, visit www.rand.org/t/RR2150

Library of Congress Cataloging-in-Publication Data is available for this publication.
ISBN: 978-0-8330-9937-2

Published by the RAND Corporation, Santa Monica, Calif.

Cover image: AP Photo/Gene J. Puskar

Preface

The RAND Corporation has a long history of research on intelligent systems. Since the 1950s, with work on chess-playing computers and the Logic Theory Machine, RAND has produced objective, evidence-based research to help inform how society can harness the benefits and manage the risks of intelligent, transformative technologies. RAND's work on autonomous and automated vehicles builds on this firm foundation. The 2009 article "Liability and Regulation of Autonomous Vehicle Technologies" and the flagship report *Autonomous Vehicle Technology: A Guide for Policymakers* in 2014 (revised in 2016) examined the policy landscape surrounding these technologies. As the technology nears readiness, policymakers today face pressing questions about how the safety of highly automated vehicles can be determined and how safe they should be before they are on the road for consumer use. The 2016 report *Driving to Safety: How Many Miles of Driving Would It Take to Demonstrate Autonomous Vehicle Reliability?* addressed the former question, showing that test-driving is not a feasible way to prove the performance of such vehicles prior to deployment. This 2017 study directly informs the latter question by assessing safety outcomes under different policies governing the introduction of highly automated vehicles. It is complemented by the report *RAND Model of Automated Vehicle Safety (MAVS): Model Documentation*, which describes in detail the model used for the analysis described in this report.

RAND Science, Technology, and Policy

This research was conducted in the RAND Science, Technology, and Policy program, which focuses primarily on the role of scientific development and technological innovation in human behavior, global and regional decisionmaking as it relates to science and technology, and the concurrent effects that science and technology have on policy analysis and policy choices.

This program is part of RAND Justice, Infrastructure, and Environment, a division of the RAND Corporation dedicated to improving policy- and decisionmaking in a wide range of policy domains, including civil and criminal justice, infrastructure development and financing, environmental policy, transportation planning and technology, immigration and border protection, public and occupational safety, energy policy, science and innovation policy, space, and telecommunications.

During the development of this report and at the time of publication, co-author Nidhi Kalra's spouse served as co-founder and president of Nuro, a machine-learning and robotics start-up company engaged in autonomous vehicle development. He previously served as a principal engineer for Google's driverless car project. Neither Kalra's spouse nor the companies he has worked for had any influence on this report.

Questions or comments about this report should be sent to the project leader, Nidhi Kalra (Nidhi_Kalra@rand.org). For more information about RAND Science, Technology, and Policy, see www.rand.org/jie/stp or contact the director at stp@rand.org.

RAND Ventures

RAND is a research organization that develops solutions to public policy challenges to help make communities throughout the world safer and more secure, healthier and more prosperous. RAND is nonprofit, nonpartisan, and committed to the public interest.

RAND Ventures is a vehicle for investing in policy solutions. Philanthropic contributions support our ability to take the long view, tackle tough and often-controversial topics, and share our findings in innovative and compelling ways. RAND's research findings and recommendations are based on data and evidence, and therefore do not necessarily reflect the policy preferences or interests of its clients, donors, or supporters.

This venture was made possible, in part, by the Zwick Impact Fund. Charles Zwick—a researcher at RAND from 1956 to 1965 who later served as both a trustee and an advisory trustee—presented RAND Ventures with $1 million and the charge to take on new and emerging policy challenges and to support top talent in their focus on these issues. Each year, RAND's president draws on this generous gift to help RAND research and outreach teams extend the impact of completed research.

Support for this project is also provided, in part, by the income earned on client-funded research, and by the generous contributions of the RAND Justice, Infrastructure, and Environment Advisory Board.

Contents

Figures and Tables

Figures

Tables

(over 15 years), more lives are cumulatively saved under the less stringent Improve10 policy than the more stringent Improve75 or Improve90 policies in nearly all conditions, and those savings can be significant—tens of thousands to hundreds of thousands of lives. The savings are greatest when HAVs under Improve10 are adopted rapidly. An Improve75 or Improve90 policy saves more lives only when HAVs introduced under Improve10 lead to a large increase in vehicle miles traveled that is not offset by a correspondingly rapid reduction in the HAV fatality rate. However, even under these conditions, the short-term life savings under the more stringent policies are relatively small (at most, approximately 3,000 lives cumulatively) and disappear over time as HAV fatality rates continue to improve under an Improve10 policy.

In the long term (over 30 years), more lives are cumulatively saved under an Improve10 policy than an Improve75 or Improve90 policy under all combinations of conditions we explored, and those savings can be very large—in some cases, more than half a million lives. The savings are largest when the introduction of HAVs under Improve75 or Improve90 is significantly delayed relative to the introduction under Improve10 because (1) the miles of real-world driving that it takes to realize significant HAV safety improvements is large and (2) the same improvement cannot be achieved equally quickly in laboratory or simulation settings. Savings are smallest when the opposite conditions hold.

Second, what does the evidence suggest about the conditions that lead to small costs from waiting for technology that is many times safer than human drivers? There is little reason to believe that improvement in HAV safety performance will be fast and can occur without widespread deployment, given the years already dedicated to HAV development and given that real-world driving is key to improving the technology. Indeed, there is good reason to believe that reaching significant safety improvements may take a long time and may be difficult prior to deployment.

Third, what does this imply for policies governing the introduction of HAVs for consumer use? In a utilitarian society, our findings would imply that policymakers should allow and developers should deploy HAVs when their safety performance is better than that of the average human driver. However, we do not live in a utilitarian society, and a potentially negative social response to HAV crashes may have profound implications for the technology. Instead, our findings suggest that society—including the public, policymakers, the judicial system, and the transportation industry—must balance the incidence of crashes from HAVs and non-HAVs with the social acceptability of each. The evidence in this report can help stakeholders find a middle ground of HAV performance requirements that may prove to save the most lives overall.

Acknowledgments

We would like to thank Anita Chandra, Marjory Blumenthal, and James Anderson for their advice and support from the very beginning of this work. We are enormously grateful to Constantine Samaras at Carnegie Mellon University and Steven Shladover at the University of California Berkeley's Partners for Advanced Transportation Technology program for their insightful reviews. Our analysis and report benefited greatly from their suggestions. Finally, we are grateful to Charles Zwick for his generous support of RAND through the Zwick Impact Awards, without which this report and related materials could not have been produced.

Abbreviations

ADS	automated driving system
DDT	dynamic driving task
FHWA	Federal Highway Administration
HAV	highly automated vehicle
MAVS	RAND Model of Automated Vehicle Safety
NHTSA	National Highway Traffic Safety Administration
ODD	operational design domain
RDM	Robust Decision Making
VMT	vehicle miles traveled

CHAPTER ONE

Introduction

Motor vehicle crashes are a public health crisis in the United States and around the world. In 2015, 35,092 people lost their lives in such crashes in the United States, an increase of 7.2 percent from 2014, and 2.44 million were injured, an increase of 4.5 percent from 2014 (National Highway Traffic Safety Administration [NHTSA], 2016a). And 2016 was even deadlier—with 37,461 fatalities (NHTSA, 2017). U.S. motor vehicle crashes can pose economic and social costs of more than $800 billion in a single year (Blincoe et al., 2015). Moreover, more than 90 percent of crashes involve driver-related errors (NHTSA, 2015; Dingus et al., 2016), such as driving too fast and misjudging other drivers' behaviors, as well as distraction, fatigue, and alcohol impairment.

Many are looking to *highly automated vehicles (HAVs)*—vehicles that drive themselves some or all of the time—to mitigate this crisis.[1] Such vehicles have the potential to eliminate many of the mistakes that human drivers routinely make (Anderson et al., 2016; Fagnant and Kockelman, 2015).[2] To begin with, HAVs are never drunk, distracted, or tired; these factors are involved in 29 percent, 10 percent, and 2.5 percent of all fatal crashes, respectively (NHTSA, 2016c; NHTSA, 2016d; NHTSA, 2011).[3] Their performance may also be better than human drivers because of better perception (e.g., no blind spots), better decisionmaking (e.g., more-accurate planning of complex driving maneuvers, such as lane changes at high speeds), and better execution (e.g., faster and more-precise control of steering, brakes, and acceleration).

But there is recognition that these vehicles, too, may pose risks to safety. For instance, inclement weather (Kutila et al., 2016) and complex driving environments

[1] We use the term *HAV* to refer to vehicles that fall into Levels 3, 4, and 5 of SAE International (2016)'s automated vehicle taxonomy. We elaborate on this and other definitions and discuss differences in terminology in Chapter Two.

[2] As we discuss in Chapter Two, vehicles that fall into SAE International's Level 1 and Level 2 automation may also help avoid many crashes caused by human error. Chapter Three describes how we incorporate these changes in our analysis.

[3] This does not mean that 41.5 percent of all fatal crashes are caused by these factors, because a crash may involve, but not be strictly caused by, one of these factors, and because more than one of these factors may be involved in a single crash.

pose challenges for HAVs (Shladover, 2016), as well as for human drivers, and HAVs might perform worse than human drivers in some of these situations (Gomes, 2014). There is also the potential for HAVs to pose new and different crash risks, such as coordinated or simultaneous crashes resulting from cyber attacks (Anderson et al., 2016; Petit and Shladover, 2015), or to suffer from hardware and software faults (Koopman and Wagner, 2017).

Clearly, HAVs present significant potential benefits and risks; they may reduce familiar risks from human drivers while simultaneously introducing unfamiliar risks from machines. Thus, HAV safety is a principal concern for the transportation industry, policymakers, and the public.[4] Assessing safety requires considering two issues: How should HAV safety be measured, and what threshold of safety should be required before HAVs are made publicly available? In essence, what test do HAVs have to take and what constitutes a passing grade? The answers to these questions would help policymakers set appropriate regulations, would enable the industry to develop appropriate tests for HAV performance, and would help the public have clearer expectations of HAV safety.

At this time, both questions remain unanswered. RAND research recently showed that the only proven method of testing safety—driving HAVs in real traffic conditions and observing their performance—requires too many miles of driving to be practical prior to widespread consumer use (Kalra and Paddock, 2016). Fortunately, there is much effort being put into developing and validating alternative methods, including accelerated testing on roads and in simulation (Zhao and Peng, 2017; Google Auto LLC, 2016) and testing for behavioral competency at closed courses and proving grounds (Nowakowski et al., 2017; U.S. Department of Transportation, 2017). Based in Germany, the Pegasus Project is a key effort to draw on and integrate these and other methods into a testing and validation framework (Lemmer, 2017).

Simultaneously, effort is needed to answer the second question of how safe HAVs should be before they are allowed on the road for consumer use. This question underpins much of the debate around how and when to introduce and use the technology so that the potential risks from HAVs are minimized and the benefits maximized. Policymakers in Congress, for example, are considering revising existing federal standards and regulations that govern traditional automobiles, which would affect how and when HAVs can be sold to consumers (U.S. Senate, 2017; Roose, 2017; Fraade-Blanar and Kalra, 2017). The HAV industry is concerned with the same question (Hsu, 2017)—not only to meet potential regulatory changes but also to meet consumer expectations, to mitigate potential backlash from the public in the event of the inevitable crash, and to manage questions of liability. And, of course, consumers will need to decide whether they have enough confidence in the performance of HAVs to hop in, and many are not so sure (Abraham et al., 2017).

---

[4] For instance, Congress has held many hearings on automated vehicles, and safety of automated vehicles is consistently a priority in statements from policymakers. See, for example, Walden (2017) and Collins (2016).

From a utilitarian standpoint, it seems sensible that HAVs should be allowed on U.S. roads once they are judged safer than the average human driver so that the lives lost from road fatalities are reduced as soon as possible. Yet, under such a policy, HAVs would still cause many crashes, injuries, and fatalities—albeit fewer than their human counterparts. This may not be acceptable to society. A large body of research suggests that peoples' willingness to accept technological risk is governed by factors related not only to the actual risk but also to other characteristics (Sjöberg, 2000; Slovic and Peters, 2006; Dietvorst, Simmons, and Massey, 2014). For example, risks are more acceptable when they are voluntary (which it may not be for the many road users who will have to share the road with HAVs) and if a person can exert control over the outcomes (which is, by definition, not the case for higher levels of vehicle automation) (Starr, 1969; Fischhoff et al., 1978; Otway and von Winterfeldt, 1982; Slovic, 1987, 2000; Dietvorst, Simmons, and Massey, 2016).

There is also a view that humans accept mistakes from other humans because they have an empathy that is not felt for machines. As Gill Pratt of Toyota Research Institute observes, "Society tolerates a significant amount of human error on our roads. We are, after all, only human. On the other hand, we expect machines to perform much better. . . . Humans have shown nearly zero tolerance for injury or death caused by flaws in a machine" (Pratt, 2017). Yet waiting for nearly perfect HAVs may miss opportunities to save lives. It is the very definition of allowing perfect to be the enemy of good.

Moreover, real-world driving experience may be one of the most important tools for improving HAV safety and, by extension, road safety. This is because, unlike most humans, HAVs can learn from each other's mistakes. When a human driver makes a mistake on the road, typically only that individual learns from that experience and potentially improves his or her driving habits. Other drivers are unaffected. This is not the case with HAVs. HAV developers use the driving experience of individual vehicles to improve the state of the art in HAV safety (Musk, 2015). The machine-learning algorithms that govern HAV perception, decisionmaking, and execution rely largely on driving experience to improve. Therefore, the more (and more-diverse) miles that HAVs drive, the more potential there is for improving the state of the art in their safety.

The lack of consensus on how safe HAVs should be before they are allowed on the road for consumer use reflects different values and beliefs when it comes to humans versus machines. But these values and beliefs can be informed by science and evidence. In this report, we seek to provide such evidence. We use the RAND Model of Automated Vehicle Safety (MAVS) (Kalra and Groves, 2017) to calculate and compare road fatalities under (1) a policy that allows HAVs to be deployed for consumer use when their safety performance is just 10 percent better than that of the average human driver (we call this option *Improve10*) and (2) a policy that waits to deploy HAVs only once their safety performance is either 75 percent or 90 percent better than that of the average human driver (we call these options *Improve75* and *Improve90*, respectively).

We use MAVS to answer three important questions:

1. Under what conditions are more lives saved by each policy in the short term and the long term, and how much are those savings?
2. What does the evidence suggest about the conditions that lead to small costs from waiting for technology that is many times safer than human drivers?
3. What does this imply for policies governing the introduction of HAVs for consumer use?

Importantly, the answers depend on deeply uncertain conditions that contribute to road fatalities, including when HAVs will be introduced to the marketplace, how quickly they will be adopted and diffused, how their safety performance will improve over time, and how the use and performance of non-HAVs will evolve. Therefore, rather than using the model to predict road fatalities under a single set of future conditions, we use methods for decisionmaking under deep uncertainty—specifically, Robust Decision Making (RDM) (Groves and Lempert; 2007; Lempert et al., 2003)—to estimate road fatalities under a wide range of potential future conditions and policies for marketplace introduction. We then analyze these results to address the three questions.

The remainder of this report is organized in four chapters. Chapter Two presents definitions of HAVs and non-HAVs, along with a review of the literature on projections of future road safety for such vehicles. Chapter Three presents our analytical method, and Chapter Four presents analytical results. Chapter Five describes the policy implications of these results and offers conclusions.

CHAPTER TWO

Definitions and Prior Work

In this chapter, we define what we mean by an HAV and describe how future road safety with and without HAVs has been assessed in the literature.

What Is a Highly Automated Vehicle and What Is Not?

The 2016 SAE International taxonomy of driving automation, summarized in Table 2.1, describes six levels of automation.[1] Common to Level 0 (no driving automation), Level 1 (driver assistance), and Level 2 (partial driving automation) is that the human behind the wheel is responsible for some or all of the dynamic driving task (DDT), even when driver assistance systems are engaged.[2] Common to Level 3 (conditional driving automation), Level 4 (high driving automation), and Level 5 (full driving automation) is that the vehicle is responsible for the entire DDT when the automated driving capabilities are engaged. As of 2017, no Level 3–5 vehicles are available for consumers to lease or purchase, but pilot tests are under way with trained safety drivers behind the wheel. Table 2.1 shows each level's description taken directly from the SAE International taxonomy (in italics) and our simplified interpretation.

Consistent with Federal Automated Vehicles Policy (NHTSA, 2016b), we use the term *HAV* to refer to vehicles that conform to Levels 3, 4, or 5.[3] We use the term *non-HAV* to refer to vehicles that conform to Levels 0, 1, or 2.

---

[1] Many terms have been coined to describe the variety of technologies that are transforming vehicles from human-driven to machine-driven, such as *automated, autonomous, self-driving*, and *driverless* vehicles. The terms are used differently in policy guidance, academic literature, and the media, and the SAE International (2016) taxonomy provides a useful discussion of their differences. In prior work (e.g., Anderson et al., 2016; Kalra and Paddock, 2016), RAND researchers have preferred the term *autonomous vehicle*, but we use the term *highly automated vehicle* here for greater consistency with federal policy (NHTSA, 2016b).

[2] SAE International (2016) defines the DDT as "all of the real-time operational and tactical functions required to operate a vehicle in onroad traffic, excluding the strategic functions, such as trip scheduling and selection of destinations and waypoints."

[3] Note that in the SAE International taxonomy, the term *highly automated* would apply to Level 4 vehicles specifically, but the Federal Automated Vehicles Policy uses the term *highly automated vehicle* more broadly.

Table 2.1
SAE International Levels of Driving Automation

Level	Name	Description
Driver performs part or all of the DDT (Non-HAVs)		
0	No driving automation	*The performance by the driver of the entire DDT, even when enhanced by active safety systems.* The human driver is entirely responsible for driving, even if such features as active electronic stability control are available and engaged.
1	Driver assistance	*The sustained and operational design domain (ODD)-specific execution by a driving automation system of either the lateral or the longitudinal vehicle motion control subtask of the DDT (but not both simultaneously) with the expectation that the driver performs the remainder of the DDT.*[a] The human driver is entirely responsible for driving but may be assisted by a single feature that automates steering or acceleration, such as lane-keeping and adaptive cruise control, but not both.
2	Partial driving automation	*The sustained and ODD-specific execution by a driving automation system of both the lateral and longitudinal vehicle motion control subtasks of the DDT with the expectation that the driver completes the object and event detection and response subtask and supervises the driving automation system.* The human driver is entirely responsible for driving but may be assisted by functions that automate both steering and acceleration, such as lane-keeping and adaptive cruise control; the human driver is responsible for monitoring the environment and intervening whenever needed.
Automated driving system performs the entire DDT while engaged (HAVs)		
3	Conditional driving automation	*The sustained and ODD-specific performance by an [automated driving system (ADS)] of the entire DDT with the expectation that the DDT fallback-ready user is receptive to ADS-issued requests to intervene, as well as to DDT performance-relevant system failures in other vehicle systems, and will respond appropriately.* The vehicle is entirely responsible for driving in certain conditions but may request rapid intervention from the human driver as needed.
4	High driving automation	*The sustained and ODD-specific performance by an ADS of the entire DDT and DDT fallback without any expectation that a user will respond to a request to intervene.* The vehicle is entirely responsible for driving in certain conditions and will not request intervention from the human driver.
5	Full driving automation	*The sustained and unconditional (i.e., not ODD-specific) performance by an ADS of the entire DDT and DDT fallback without any expectation that a user will respond to a request to intervene.* The vehicle is entirely responsible for driving under all conditions and will not request intervention from anyone in the vehicle. Such vehicles may have no occupants at all.

SOURCE: SAE International, 2016.

NOTE: Below each italicized description taken directly from the SAE International taxonomy (2016, Table 2) is our simplified interpretation.

[a] SAE International (2016) defines an ODD as "the specific conditions under which a given driving automation system or feature thereof is designed to function."

How Has Future Road Safety Been Assessed in the Literature?

There are many estimates of the benefits of different types of advanced driver assistance systems or crash avoidance systems, individually and in combination (Gordon et al., 2010; Funke et al., 2011; Perez et al., 2011; Jermakian, 2011; Harper, Hendrickson, and Samaras, 2016). Vehicles equipped with these technologies are usually classified as having Level 1 or Level 2 automation according to SAE International's taxonomy and are considered non-HAVs. As one example, Funke et al. (2011) summarizes the safety potential of four crash avoidance technologies as the product of the size of the crash problem in the entire U.S. fleet (e.g., the number of annual crashes related to lane departure) and the fraction of such crashes that could be mitigated by the technology (e.g., from a lane departure warning system). There are also efforts to estimate the benefits of connected vehicle technologies, in which on-board applications use communication with other vehicles or infrastructure to improve safety—for example, for coordinating vehicle movement through an intersection (Najm, Toma, and Brewer, 2013; Eccles et al., 2012).

Rau, Yanagisawa, and Najm (2015) describes a method for identifying the types and potential number of current crashes that could be mitigated by technologies between Level 2 and Level 5 autonomy. Li and Kockelman (2016) draws on this methodology to estimate the safety benefits of a variety of connected vehicle and Level 1 and Level 2 automated vehicle technologies, assuming widespread adoption of those technologies. Going a step further than prior work, Li and Kockelman (2016) estimates both the types and severity of crashes that could be avoided by each type of technology, as well as the economic benefit of those savings. In recognition of the uncertainty in the technology performance, the authors assess benefits under three scenarios of technology effectiveness.

There are fewer estimates of the safety benefits of HAVs, and there is no consensus yet among those estimates (Winkle, 2015). Fagnant and Kockelman (2015) calculate the societal benefits of Level 4 and Level 5 HAVs across a variety of benefit categories, including safety.[4] Drawing on the findings of the National Motor Vehicle Crash Causation Survey, which found that human error accounts for 93 percent of today's crashes (NHTSA, 2008), Fagnant and Kockelman assume in their calculations that HAVs reduce crash and injury rates by 50 percent at the 10-percent market penetration rate and by 90 percent at the 90-percent market penetration rate. In contrast, the Casualty Actuarial Society's Automated Vehicles Task Force recently evaluated the findings of the National Motor Vehicle Crash Causation Survey in the context of HAVs. The task force's study found that HAVs could address about half of the accidents, while "49% of accidents contain at least one limiting factor that could disable [HAV] technology or reduce its effectiveness" (Casualty Actuarial Society, 2014).

---

[4] The paper does not explicitly define what levels of autonomy the authors include in their calculations, but we infer that they refer to Level 4 and Level 5 autonomy, not Level 3.

This literature provides important insights into how different non-HAV and HAV technologies could mitigate today's crashes. However, these insights have not yet been used to understand how safety effects might play out over time—because different technologies are adopted in different time frames, and the performance of the technologies changes as they are deployed. It is difficult to use these estimates to make such projections for two key reasons. First, the estimates generally focus on how technologies could mitigate the types of crashes that human drivers currently cause, but they overlook important ways in which new technologies could add to crashes. This could occur if technology erodes human drivers' skills or attention, technology is vulnerable to cybersecurity failures that lead to new types of crashes, or HAVs simply perform worse than human drivers, even initially (Kalra, 2017).[5] As the Casualty Actuarial Society notes, "The safety of automated vehicles should not be determined by today's standards; things that cause accidents today may or may not cause accidents in an automated vehicle era" (2014, p. 1). New safety risks are difficult to anticipate, making the full effect of many new technologies deeply uncertain.

Second, these existing estimates also compare the marginal benefits of a technology with the safety performance of current vehicles and drivers. However, the benefit of a vehicle with a particular technology at some point in the future is more correctly estimated when compared with the performance of future vehicles without that technology at that same future time, rather than with vehicles in current conditions.

The history of airbags illustrates these issues (Anderson et al., 2016; Houston and Richardson, 2000). When airbags were first introduced in the 1970s, they were designed to protect an unbelted adult male passenger and envisioned as an alternative rather than a supplement to seat belts, which were then used infrequently. Estimates of future safety benefits made at that time were based on this use case and ultimately proved to be overblown by an order of magnitude—in large part because, by the time airbags were widely deployed, seat belt use was also widespread, so the marginal benefit of airbags was much less than anticipated. Moreover, while airbags still saved many lives, the force needed to protect an unbelted adult male injured and killed many passengers of smaller stature (such as women and children) who might have otherwise survived the crash had airbags not deployed. These crashes led to improvements in airbag technology but also showed that airbags introduced new crash risks even as they mitigated existing risks.

In sum, the long-term evolution of road safety is important to understand and yet complex, deeply uncertain, and difficult to predict. This work fills a gap in the existing literature by using a simple modeling platform to explore the safety impact of HAVs under different policies and conditions.

---

[5] Complicating matters, as Kalra and Paddock (2016) argues, there is no currently accepted method of assessing the safety of HAVs with statistical confidence prior to making them available for widespread use. Therefore, it is possible that stakeholders may simply not know how safe the technology is.

CHAPTER THREE

Methods

The short- and long-term safety outcomes of different HAV policies will depend on the evolution of many factors, such as use and safety of non-HAVs over time; the timing, rate, and extent of HAV adoption and diffusion throughout the fleet; and the initial safety of HAVs and how much and how quickly it improves. Accurately predicting safety outcomes is fraught with complications because such factors are *deeply uncertain*, meaning there is no consensus about how they will evolve and any prediction is likely to be wrong given the disruptive nature of the technology. Therefore, such predictions may ultimately not be helpful in determining which policy would lead to better safety outcomes. We therefore turn to methods for decisionmaking under deep uncertainty (Kalra et al., 2014)—specifically, RDM (Lempert, Popper, and Bankes, 2003; Groves and Lempert, 2007).

The remainder of this chapter presents our methodology and experimental design in greater detail. We first provide an overview of RDM and then MAVS. Later, we define our policies and explain how we account for uncertainties that govern the performance of each policy.

Overview of Robust Decision Making

Quantitative analysis is often indispensable for making sound policy choices. Typically, these methods use a "predict-then-act" approach: Analysts assemble available evidence into best-estimate assumptions or predictions and then use models and tools to suggest the best strategy given these predictions. Such analyses are useful in answering the question, Which policy options best meet our goals *given our beliefs about the future*? These methods, which include probabilistic risk analysis, work well when the predictions are accurate and noncontroversial (Lempert, Popper, and Bankes, 2003; Kalra et al., 2014; Lempert and Kalra, 2011).

However, disruptive technologies (such as HAVs), by definition, do not lend themselves to credible prediction-making. As noted, the short- and long-term safety outcomes of different HAV introduction policies will depend on the evolution of many deeply uncertain factors. Traditional methods prove brittle in the face of the deep uncertainties. Disagreements about future predictions can lead to gridlock among stakeholders. Worse, decisions tailored to one set of assumptions often prove inadequate or even harmful if another future comes to pass.

Many methods have been developed over the past half-century to help policymakers manage deep uncertainties and make choices that are robust to the unpredictable future. RDM, in particular, is designed to help manage deep uncertainty by helping develop policies that are robust—that is, that satisfy decisionmakers' objectives in many plausible futures rather than being optimal in any single best estimate of the future (Lempert et al., 2013).

RDM rests on a simple concept. Rather than using models and data to assess policies under a single set of assumptions, RDM runs models over hundreds or thousands of different sets of assumptions to describe how plans perform in many plausible conditions. Unlike, for example, Monte-Carlo analysis, which attaches probabilities to those assumptions to estimate expected outcomes, RDM uses simulations to stress test strategies. RDM draws from both scenario planning and probabilistic risk analysis to ask which policies reduce risk over which range of assumptions, inquiring, for example, what assumptions would need to be true for us to reject option A and instead choose option B.

By embracing many plausible sets of assumptions or futures, RDM can help reduce overconfidence and the deleterious impacts of surprise, can systematically include imprecise information in the analysis, and can help decisionmakers and stakeholders who have differing expectations about the future nonetheless reach consensus on action. In essence, RDM helps plan for the future without first predicting it. RDM has been applied to water resource management (Groves, Davis, et al., 2008; Groves, Fischbach, et al., 2013), flood risk management (Fischbach et al., 2017), terrorism risk insurance (Dixon et al., 2007), energy investments (Popper et al., 2009), and other sectors.

In this analysis, we use RDM and MAVS (Kalra and Groves, 2017) to

- generate a wide range of plausible future conditions that would shape HAV safety outcomes, without ascribing likelihoods to those futures
- assess fatalities over time under different HAV introduction policies across those futures
- identify the set of future conditions that lead to more life savings under each policy in the short term (2020–2035) and over the long term (2020–2050)
- assess the plausibility of those conditions to determine whether one policy is more robust—that is, more likely to yield life savings despite deep uncertainties.

The results, which we discuss in Chapter Four, help inform whether it is better to wait for near-perfect performance before HAVs are allowed on the road for consumer use or better to deploy HAVs when their safety performance is only moderately better than that of the average human driver.

Overview of the Model and the Experimental Design

MAVS is a model that estimates traffic fatalities over time in a *baseline* future without HAVs and an *alternative* future with HAVs.[1] The calculations are based on a variety of factors, including

- the change in safety performance of non-HAVs over time
- travel demand for non-HAVs over time
- the timing of HAV market introduction and the rate and level of diffusion and use over time
- the safety performance of HAVs over time.

Figure 3.1 diagrams the key inputs and outputs of MAVS, including fatality rate, year of introduction, and vehicle miles traveled (VMT), among others.

Figure 3.1
MAVS Inputs and Outputs

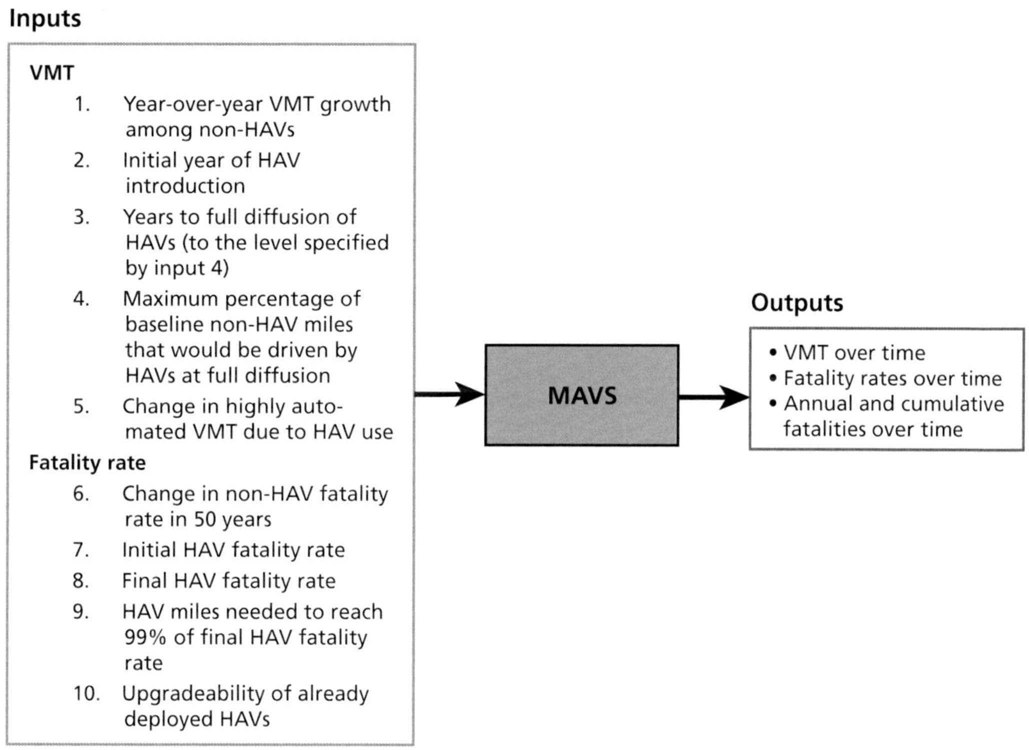

Inputs

VMT

1. Year-over-year VMT growth among non-HAVs
2. Initial year of HAV introduction
3. Years to full diffusion of HAVs (to the level specified by input 4)
4. Maximum percentage of baseline non-HAV miles that would be driven by HAVs at full diffusion
5. Change in highly automated VMT due to HAV use

Fatality rate

6. Change in non-HAV fatality rate in 50 years
7. Initial HAV fatality rate
8. Final HAV fatality rate
9. HAV miles needed to reach 99% of final HAV fatality rate
10. Upgradeability of already deployed HAVs

→ **MAVS** →

Outputs

- VMT over time
- Fatality rates over time
- Annual and cumulative fatalities over time

RAND *RR2051-3.1*

---

[1] MAVS can be configured to evaluate crashes, injuries, property damage, economic costs, or other safety measures. For simplicity and because of the particular attention paid to road deaths, we measure safety by the number of fatalities and the fatality rate.

By defining different combinations of input values, we use MAVS to model different policies under many plausible future conditions. Table 3.1 summarizes the experimental design used for this analysis. It lists each of the ten inputs to MAVS shown in Figure 3.1 (numbered for convenience) and notes how each is defined under the policies in our analysis. In this report, we define the *benchmark* fatality rate as the current fatality rate of human drivers in the United States (1.12 fatalities per 100 million miles in 2015; see NHTSA, 2016a).[2]

Table 3.1
Summary of the Experimental Design

MAVS Input	Improve10 Policy	Improve75 and Improve90 Policies
VMT		
1. Year-over-year VMT growth among non-HAVs	Uncertain; 0.4%–1.8%	
2. Initial year of HAV introduction	Uncertain; constant 2020	Uncertain; affected by fatality rates in Improve10 and an uncertain delay of 0–15 years in the introduction of HAVs
3. Years to full diffusion of HAVs (to the level specified by input 4)	Uncertain; 20–50 years	Uncertain; defined by an uncertain amount of accelerated diffusion relative to Improve10
4. Maximum percentage of baseline non-HAV miles that would be driven by HAVs at full diffusion	Uncertain; 50%–100%	
5. Change in highly automated VMT due to HAV use	Uncertain; -50%–100%	
Fatality rate		
6. Change in non-HAV fatality rate in 50 years	Uncertain; 50%–110% of the benchmark fatality rate	
7. Initial HAV fatality rate	Defined by policy as 90% of the benchmark fatality rate	Defined by policy as 25% (Improve75) or 10% (Improve90) of the benchmark fatality rate
8. Final HAV fatality rate	Uncertain; constant, defined by initial fatality rates under Improve75 and Improve90	
9. HAV miles needed to reach 99% of final HAV fatality rate	Uncertain; 100 million to 10 trillion	Not applicable
10. Upgradeability of already deployed HAVs	Uncertain; 0–1	Not applicable

[2] In October 2017, just prior to this report's publication, NHTSA released traffic safety data for 2016 and revised estimates for 2015. NHTSA reports that, in 2016, the fatality rate increased still further to 1.18 fatalities per 100 million VMT from a (revised) rate of 1.15 fatalities per 100 million VMT in 2015 (NHTSA, 2017). The analysis in this report is based on the earlier 2015 estimate of 1.12 fatalities per 100 million VMT.

One MAVS input (input 7) defines the HAV policy, while most of the others are uncertain and govern how each policy may perform. For each uncertain input, we note in the table whether the input is treated as a constant (e.g., input 2), explored over the stated range (e.g., input 1), or defined by other factors and uncertainties (e.g., input 3 under the Improve75 and Improve90 policies).

Consistent with RDM, our experimental design includes a wide range of plausible values for the MAVS parameters that will shape HAV safety but does not ascribe likelihoods to any particular set of values or futures. Therefore, these ranges may include values that are not viewed as plausible by all stakeholders. However, the RDM method is not highly sensitive to extended ranges, because the analytics focuses on identifying thresholds that favor different policies rather than on finding optimal policies.

The next step in our analysis is to define 500 plausible futures.[3] A *future* is a specific and unique combination of the uncertain, nonconstant inputs into MAVS, and we generate the ensemble using a Latin Hypercube sampling procedure.[4] We evaluate the HAV introduction policies under each future and save the results in a database for analysis, as described in Chapter Four. The following sections elaborate on the introduction policies and ranges of values used to represent the uncertainties and to generate the ensemble of case runs.

Defining Highly Automated Vehicle Introduction Policies

We have configured MAVs to examine the long-term safety outcomes of HAV policies that differ by the level of safety performance HAVs must attain before they are allowed on U.S. roads for consumer use. Therefore, each policy defines MAVS input 7 (initial HAV fatality rate).

The first policy, Improve10, allows HAVs to be deployed once their fatality rate is one fatality per 100 million VMT, or 10 percent better than the benchmark rate (1.12 fatalities per 100 million miles, as noted earlier).[5]

The second policy we examined allows HAVs to be deployed only once their performance is several times better than that of human drivers. We define two variations, given that it is uncertain how safe HAVs can ultimately become and how much toler-

[3] The number of futures is arbitrary, but significantly fewer futures may lead to insufficient exploration of the experimental design space, while significantly more futures would not necessarily add more insight yet may be more difficult to calculate and visualize in the results.

[4] A Latin Hypercube Sampling procedure ensures that all variables are sampled uniformly across their entire range and that the combinations of values across the variables are randomly selected (Saltelli, Chan, and Scott, 2000).

[5] Our choice of a 10-percent improvement is arbitrary: Any technology that reduces fatality rates even slightly (e.g., 1 percent) relative to human drivers would be better than average. One practical reason to use a modest difference over a very small one, however, is that it becomes more feasible to detect and verify such a difference in performance (Kalra and Paddock, 2016).

ance stakeholders have for imperfection. The first variation, Improve75, allows HAVs only once their fatality rate is 0.28 per 100 million VMT, representing a 75-percent improvement over the benchmark rate. The second variation, Improve90, is more stringent, allowing HAVs only once their fatality rate is 0.11 fatalities per 100 million VMT, representing a 90-percent improvement over the benchmark rate.

Accounting for Uncertainty in a Baseline Future Without Highly Automated Vehicles

MAVS first calculates annual VMT, annual fatality rates, and (from these factors) annual fatalities in a future that has no HAVs. This serves as a baseline from which a future with HAVs deviates. Annual VMT is calculated based on an uncertain year-over-year growth in VMT (MAVS input 1 in Table 3.1). In its 2017 projections, the Federal Highway Administration (FHWA) forecasts that, from 2015 to 2045, total VMT could grow in the range of 0.66 percent to 0.89 percent annually (FWHA, 2017). The FHWA bounds are uncertain and have changed significantly year to year. In 2016, for example, the forecasted increase in VMT ranged from 0.53 percent to 0.65 percent, meaning that the high end of FHWA's 2016 projection (0.65 percent) was lower than the low end of its 2017 projection (0.66 percent) (FHWA, 2016). The Energy Information Administration's Annual Energy Outlook has similar projections and variation across years. The 2017 report projects a growth rate of 0.7 percent for light-duty vehicles out to 2050 and greater annual increases for commercial trucks (Energy Information Administration, 2017a). In recent decades, the projections have ranged significantly, from 0.7-percent annual growth to 1.8-percent annual growth for certain periods (Energy Information Administration, 2006).[6] This variability speaks to the deep uncertainty surrounding long-term future VMT. Based on this uncertainty and related literature, in our experiments, we explore over a large range: 0.4 percent to 1.8 percent.

The next uncertainty is the fatality rate among non-HAVs over time (MAVS input 6). While road fatality rates have declined significantly since the 1950s, there has been a plateau over the past decade and an increase in recent years (Bureau of Transportation Statistics, 2016). Based on the estimates of the safety benefits of driver assistance systems described in Chapter Two, we allow for up to a 50-percent decrease in the long-term fatality rates of non-HAVs. However, it is also possible that fatality rates could increase if the safety benefit of these technologies is outpaced by a decline in driver attentiveness and skill (which will still be essential for non-HAV driving) or other maladaptive behaviors (Milakis, van Arem, and van Wee, 2017). Therefore, we allow for a 10-percent increase in non-HAV fatality rates relative to the benchmark rate.

---

[6] These projections may include potential changes to VMT from HAV use, which we seek to exclude in the baseline. FHWA projects a range of growth rates in order to account for a variety of uncertainties, including future economic growth, vehicle use and ownership, and technology (FHWA, 2017).

Accounting for Uncertainty Under an Improve10 Policy

MAVS next calculates HAV and non-HAV VMT and fatality rates in a future in which HAVs are introduced under an Improve10 policy (that is, HAVs are 10 percent safer than the average human driver). These calculations consider the diffusion of HAVs, represented over time as the percentage of VMT that ultimately would be attributed to HAVs and the time in which it would take to reach *full diffusion* (that is, some ultimate level of saturation throughout the fleet, as defined by input 4).

Uncertainties Governing Highly Automated Vehicle Miles Traveled Under Improve10

Annual highly automated VMT under Improve10 first depends on timing: When will HAVs be introduced (MAVS input 2) and when will they be fully diffused through the fleet (MAVS input 3)? HAV introduction is uncertain because we do not know when HAVs will be safer than humans. For simplicity, we have chosen a near-term fixed year of introduction—2020—relative to which all outcomes in both policies are calculated. (We could have also chosen to call this year 0, but 2020 is more conceptually meaningful and also consistent with statements from automakers about when they anticipate HAVs could hit the roads (Ross, 2017; Ford Motor Company, 2016).

We also do not know when HAVs will reach full diffusion. If the diffusion of other automotive technology is any indication (Jutila, 1987; DeCicco, 2010; Highway Loss Data Institute, 2012), the full diffusion of HAVs will take at least a few decades (Litman, 2017; Bansal and Kockelman, 2017). There are some suggestions that it could be very fast, particularly in urban areas (Boston Consulting Group, 2017). To be inclusive, we consider a range of 20–50 years to full diffusion under an Improve10 policy.

HAV VMT is secondly governed by the extent of HAV use, which we measure at the point of full diffusion. To begin with, some percentage of non-HAV driving in the baseline future will ultimately be replaced by HAVs (MAVS input 4). This is deeply uncertain, and we explore anywhere from 50 percent to 100 percent of baseline miles. (The miles that are not replaced define the VMT of non-HAVs.) HAVs are also expected by many to increase the demand for transportation by reducing its costs and by increasing opportunities for automotive travel for those who currently do not drive or use other modes (Anderson et al., 2016; Harper et al., 2016; U.S. Department of Energy, 2017). Zero-occupancy vehicles are a particular concern, as is an increase in VMT from delivery of goods. Yet, there could also be some decrease in VMT if shared HAVs in particular are widely used (Energy Information Administration, 2017b).

MAVS models a change in VMT as a result of HAV use (MAVS input 5). This input asks, for each mile of human driving that is replaced by highly automated driving, how large is the replaced mile? That is, how many highly automated miles replace the human-driven mile? For instance, HAVs may lead workers to accept longer commutes because the driving time could be spent on other desired activities, such as reading or watching media. This could result in more miles traveled overall; for example, one human-driven mile could be replaced by 1.25 highly automated miles, reflecting a change of 25 percent in VMT as a result of HAVs. Alternatively, one human-driven mile could be replaced by 0.75 highly automated miles, reflecting a change of –25 percent in VMT as a result of HAVs. Given the deep uncertainties, we consider a range of effect anywhere from –50 percent to 100 percent, reflecting a halving to a doubling of VMT. This range comfortably includes all of the projections of VMT change we found in the literature.

Uncertainties Governing Highly Automated Vehicle Fatality Rates Under Improve10

The initial fatality rate of HAVs under Improve10 is defined by the policy itself: HAVs are 10 percent better than the benchmark rate when introduced into the market. Over time, this rate improves as the state of the art in HAV safety improves and as the existing HAV fleet is upgraded to keep up. Although the best possible HAV safety rate is, in reality, uncertain, we treat it as constant, linking it to the notion of "near perfect" under the Improve75 and Improve90 policies. That is, under Improve10, the state of the art in HAVs changes from a 10-percent improvement over the benchmark rate to either a 75-percent or a 90-percent improvement over the benchmark rate (MAVS input 8), depending on the policy with which it is being compared.

The rate of this improvement is also uncertain. MAVS defines it in terms of the cumulative number of miles of postdeployment HAV driving needed to improve from an initial to a final fatality rate (MAVS input 9). There is little real-world evidence that would help bound this number. As of December 2016, Tesla reportedly had logged 1.3 billion miles of real-world driving data from vehicles equipped with its Autopilot hardware (Tesla, 2016) and, as of May 2017, Waymo had logged 3 million miles of driving (Waymo, undated). There is not reason to believe that either technology is many times better than human drivers, and there are not credible data about the rate at which performance is improving. Under an Improve10 policy, we explore over a large range of improvement rates, from 100 million to 10 trillion miles of cumulative postdeployment driving. Because we are concerned with orders of magnitude of miles needed to improve, we sample in this space logarithmically (rather than linearly as we do for other uncertainties). In other words, we take the same number of samples between 100 million and 1 billion miles as we do between 1 billion and 10 billion miles.

Finally, while the state-of-the-art HAV fatality rate may improve at a particular rate, the extent to which the improvement can be diffused throughout the already-deployed fleet is uncertain. This is determined by the upgradeability of the existing fleet, either through hardware and software updates to existing vehicles or through vehicle turnover. MAVS uses an upgradeability term (MAVS input 10). A lower bound of 0 represents no upgradeability of existing vehicles; that is, once an HAV is on the road, it remains on the road with the performance it had when it was first introduced. An upper bound of 1 represents perfect upgradeability of existing vehicles; that is, every HAV always performs at the state of the art level of safety. Neither of these is true to reality, and upgradeability is explored between these bounds under an Improve10 policy.

Accounting for Uncertainty Under Improve75 and Improve90 Policies

For comparison, MAVS calculates HAV and non-HAV miles traveled and fatality rates in a future in which HAVs are introduced under policies requiring them to be 75 percent and 90 percent safer than current vehicles.

Uncertainties Governing Highly Automated Vehicle Miles Traveled Under Improve75 and Improve90

As with Improve10, annual highly automated VMT under Improve75 and Improve90 depends on the timing of introduction, the timing of full diffusion, and the extent of use. Introduction (MAVS input 2) depends on when HAVs will reach the target safety performance of 75- or 90-percent improvement. Note that under an Improve10 policy, HAV developers have all the tools to improve HAVs as they would under Improve75 or Improve90, plus the additional tool of postdeployment learning, which is not available under these latter policies. This means that HAVs under Improve10 would reach target safety performance at the same time as or sooner than under Improve75 or Improve90. Thus, the soonest HAVs can be introduced under these latter policies is when they also reach target safety performance under Improve10. This is an endogenous variable in our model, determined by highly automated VMT and improvement rates under Improve10. In addition, however, if postdeployment learning is a particularly valuable tool, HAVs under Improve75 and Improve90 may not reach target performance until sometime after they would under an Improve10 policy. This delay is treated as exogenous and ranges from 0 to 15 years. A delay of zero years signifies that postdeployment learning has no value over other development methods (simulation, test driving, etc.), while a delay of 15 years suggests that postdeployment learning has significant value. The introduction year is thus the sum of the year in which Improve10 reaches target safety performance and an additional delay caused if improvement proves difficult without postdeployment learning.

We also do not know how long it will take for HAVs under Improve75 and Improve90 to be fully diffused (MAVS input 3), but we can define the duration relative to the diffusion time specified in the same future for Improve10. It is reasonable that HAVs under Improve75 and Improve90 would not take *more years* to reach full diffusion than they would under Improve10, given that HAV performance under Improve10 is not nearly as good and consumers may be more uncertain or less ready to use the technology than they would under the other policies. Therefore, the upper bound of diffusion time is the same as the diffusion time under Improve10.

Instead, HAVs under Improve75 and Improve90 might reach full diffusion in *fewer years* because there may be latent demand for the technology because of a potential delay in introduction. However, we would not expect the HAVs to reach full diffusion *before* they would under Improve10 because, by the time HAVs are introduced under Improve75 or Improve90, they have target safety performance under Improve10 as well. Therefore, the lower bound of diffusion time is the number of years until HAVs under Improve10 have been fully diffused, if full diffusion has not already occurred. Otherwise, it is one year. We specify a weight between 0 and 1 that explores between these bounds.

The uncertainties that govern HAV use (MAVS inputs 4 and 5) are defined the same way under all policies. Their values are also the same for all policies in any particular future.

Uncertainties Governing Highly Automated Vehicle Fatality Rates Under Improve75 and Improve90

The initial fatality rate of HAVs under Improve75 and Improve90 is defined by each policy. As noted earlier, the best possible HAV fatality rate is uncertain, but, for simplicity, we assume that the very definition of "near perfection" approaches a practical limit to HAV performance. Thus, HAVs under these policies do not improve over time. The final HAV fatality rate (MAVS input 8) is defined as the same as the initial HAV fatality rate (75 percent or 90 percent improved over benchmark), and the learning rate and upgradeability (MAVS inputs 9 and 10) do not apply to HAVs under these policies.

CHAPTER FOUR

Analytical Results

In this chapter, we use the model and methods described in Chapter Three to answer the first question posed in the introduction: Under what conditions are more lives saved by each policy in the short term and the long term, and how much are those savings? We first examine the results in the short term and then in the long term.

Under What Conditions Are More Lives Saved by Each Policy in the Short Term, and How Large Are Those Savings?

Figure 4.1 shows the difference in cumulative fatalities between Improve10 and Improve75 (Panel A) and between Improve10 and Improve90 (Panel B) across the ensemble of 500 cases in the short term, measured 15 years after initial deployment of HAVs under an Improve10 policy. Positive values indicate cases in which Improve10 saves more lives cumulatively (shown in light blue when the savings are less than 50,000 and in dark blue when savings exceed 50,000), while negative values indicate cases in which Improve75 or Improve90 saves more lives cumulatively (shown in red).

These results show that an Improve10 policy saves more lives than the other policies under nearly every combination of conditions examined (476 of 500, or 95 percent, of all cases when compared with Improve75; and 484 of 500, or 97 percent, of all cases when compared with Improve90).[1] When compared with the savings under Improve75, the cumulative savings under Improve10 can reach nearly 200,000 lives; when compared with the savings under Improve90, Improve10 savings can exceed 200,000 lives. More lives are saved in the latter comparison than the former because (1) the introduction of HAVs under Improve90 is delayed more than under Improve75, creating additional opportunity for HAVs introduced under Improve10 to save lives, and (2) HAVs introduced under Improve10 can achieve lower fatality rates under Improve90 than under Improve75, creating greater means for such HAVs to save lives.

[1] This should not be interpreted as an indicator of the likelihood of different life-saving outcomes, because the cases are not generated from an underlying probability distribution of inputs. The deep uncertainties preclude reliable probabilistic forecasting of uncertainties and outcomes.

Figure 4.1
Ensemble Difference in Cumulative Lives Saved over 15 Years for 500 Cases

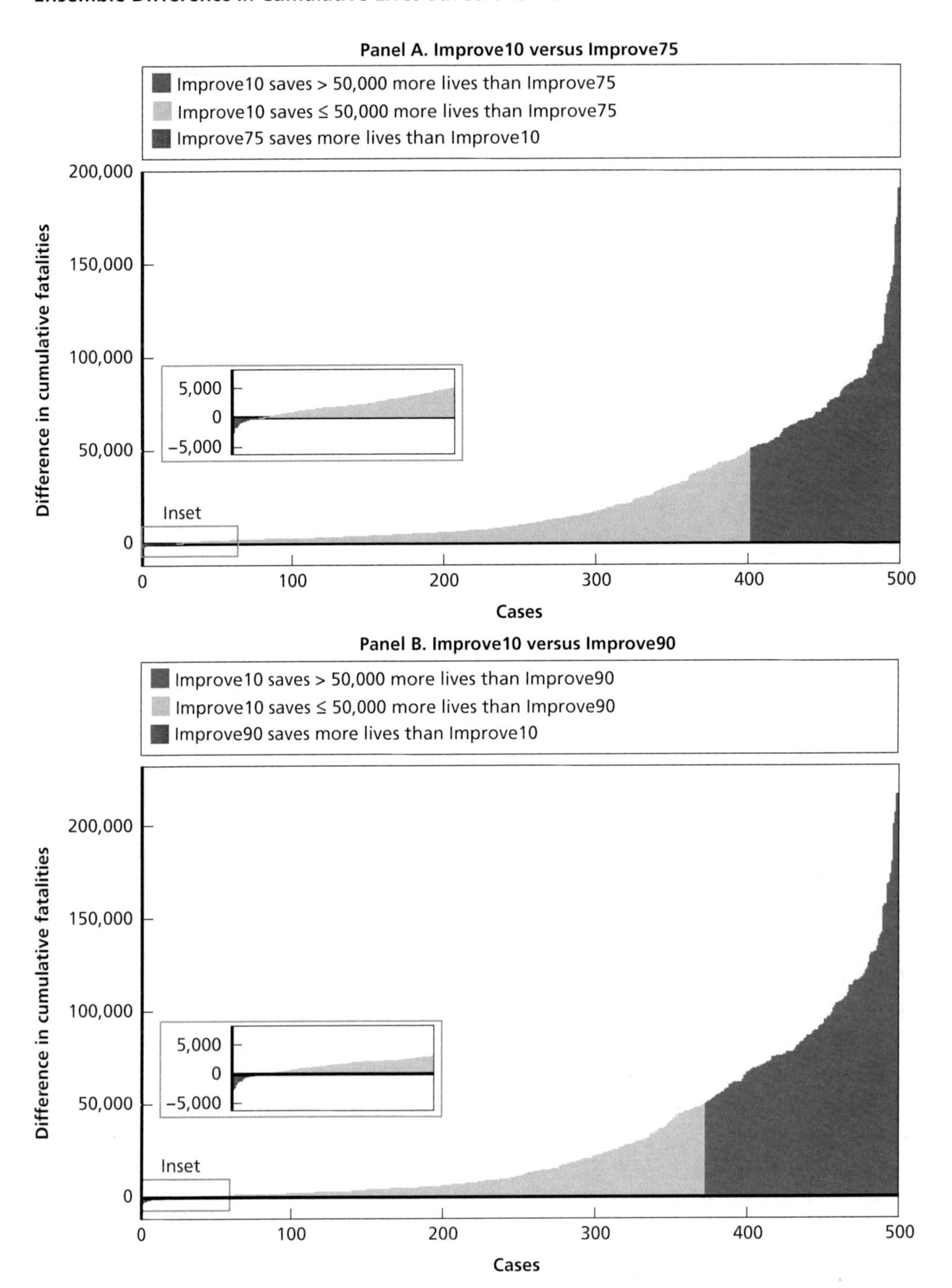

RAND *RR2051-4.1*

Figure 4.2 shows that, in the short term, the difference in cumulative fatalities (vertical axis) is driven largely by how long it takes for HAVs to reach full diffusion under an Improve10 policy (horizontal axis). This is to be expected: Faster diffusion means that more miles can be driven sooner by safe HAV technology, leading to greater opportunities for fatality differences between the two policies. Conversely, if diffusion takes many decades, few VMT are driven by HAVs in either policy, resulting in a smaller potential difference. Each dot in the figure represents one of the 500 cases.

There are a few conditions in which an Improve75 or Improve90 policy saves more lives than an Improve10 policy in the 15-year time frame. By 2035, the most cumulative lives saved by the larger improvement policies compared with Improve10 is approximately 3,000, far fewer than the tens of thousands and sometimes hundreds of thousands of lives that Improve10 saves under many of the conditions we examined.

Figure 4.2
Difference in Improve10 and Improve75 Cumulative Fatalities over 15 Years by Number of Years to Full Diffusion Under an Improve10 Policy

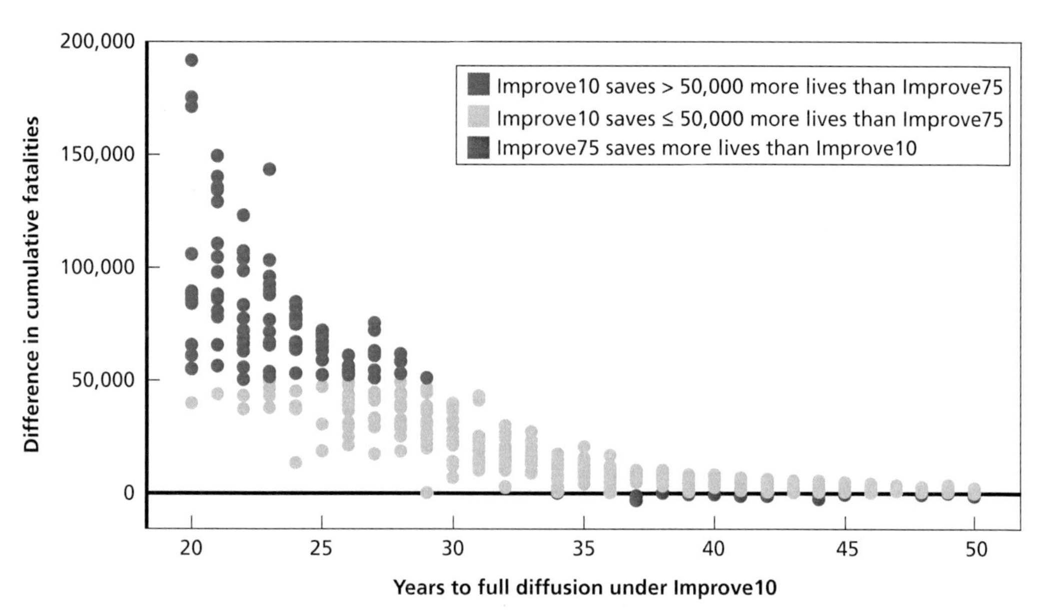

RAND *RR2150-4.2*

The savings of up to 3,000 lives under an Improve75 or Improve90 policy occur in futures in which the fatality rate of HAVs introduced in the early years of an Improve10 policy is not yet low enough to offset an increase in VMT resulting from HAV use. The slow improvement in the HAV fatality rate under Improve10 is, in turn, the result of a slow learning rate and relatively slow diffusion of the technology. Figure 4.3 shows this effect. Each mark in this figure is a future (or case) in which Improve75 has fewer cumulative fatalities than Improve10 by 2035. The size of each mark corresponds to the difference in cumulative fatalities. The figure shows that all of these futures have a 23 percent or larger increase in VMT as a result of HAV use (horizontal axis) and require 1 trillion miles or more of driving to achieve a 75-percent improvement in fatality rates over the benchmark rate.

We can explore the effects of these factors in a single future. Case 117 shows that Improve75 saves 3,200 more lives than Improve10 does in the first 15 years after HAV introduction. The uncertainty parameter values in this case are shown in Table 4.1, and the next three figures show the impact of these values on annual VMT (Figure 4.4), fatality rate (Figure 4.5), and annual fatalities (Figure 4.6).

Figure 4.3
Factors That Result in Fewer Cumulative Fatalities Under Improve75 Than Under Improve10 over 15 Years

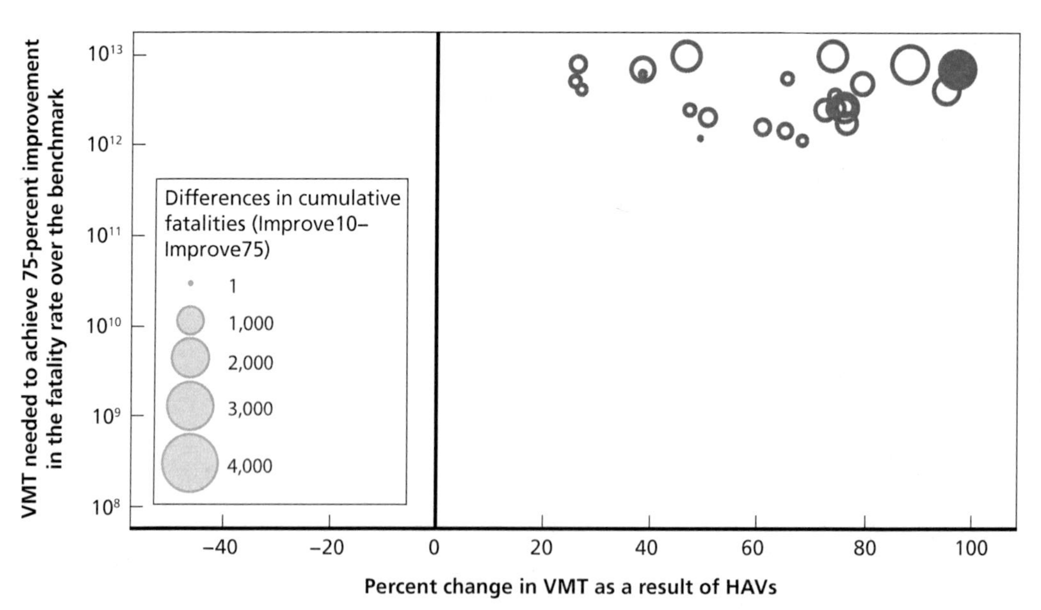

RAND *RR2150-4.3*

Table 4.1
Uncertainty Parameter Values in Case 117

MAVS Input	Improve10 or Improve 75	Uncertainty	Value	Range Evaluated
1	Both	Year-over-year VMT growth among non-HAVs	0.56%	0.4%–1.8%
2	Improve75	Years of delay in introducing HAVs under Improve75[a] (used to calculate Initial year of HAV introduction)	8	0–15
3	Improve10	Years to full diffusion of HAVs under Improve10	37	20–50
3	Improve75	Increase in pace of HAV adoption under Improve75[a] (used to calculate years to full diffusion of HAVs)	0.59	0–1
4	Both	Maximum percentage of baseline non-HAV miles that would be driven by HAVs at full diffusion	60%	50%–100%
5	Both	Change in highly automated VMT as a result of HAV use	97%	–50%–100%
6	Both	Change in non-HAV fatality rate in 50 years	0.81	0.5–1.1
9	Improve10	HAV miles needed to reach 99% of final HAV fatality rate	6.9 trillion	100 million–10 trillion
10	Improve10	Upgradeability of already deployed HAVs	0.58	0–1

[a] As described in Chapter Three, these uncertainty parameters are not directly used in MAVS but are instead used to calculate the indicated MAVS input.

The row for the "change in highly automated VMT as a result of HAV use" in Table 4.1 shows that, in this case, the diffusion of HAVs significantly increases VMT in this case.[2] The effect of this increase can be seen in Figure 4.4, in which the use of HAVs under Improve10 (in blue) causes a significant increase in VMT 15 years after HAV introduction compared with Improve75 (in red), in which HAVs are not yet introduced, so no increase occurs.

The row for "HAV miles needed to reach 99% of final HAV fatality rate" in Table 4.1 shows that it takes 6.9 trillion miles of postdeployment driving for state-of-the-art HAVs under Improve10 to improve the annual fatality rate by 90 percent over benchmark. A slower pace of diffusion (35 years) means that these miles are gained slowly. The effect is evident in Figure 4.5, when 15 years after HAV introduction and 1.6 trillion miles of cumulative HAV driving, the fatality rate of state-of-the-art HAVs (in orange) under an Improve10 policy has improved 50 percent over benchmark, but it still remains below the 75-percent improvement that constitutes best performance under this policy. The effective improvement across the entire Improve10 fleet (HAVs and non-HAVs, shown in blue) is approximately 15 percent because most vehicles are

[2] Specifically, in this future, for each vehicle mile driven by humans in the baseline case that is driven by an HAV in a future with HAVs (Improve10 or Improve75), 0.97 additional miles are driven by HAVs as a result of increased demand for travel.

Figure 4.4
Annual VMT in Case 117

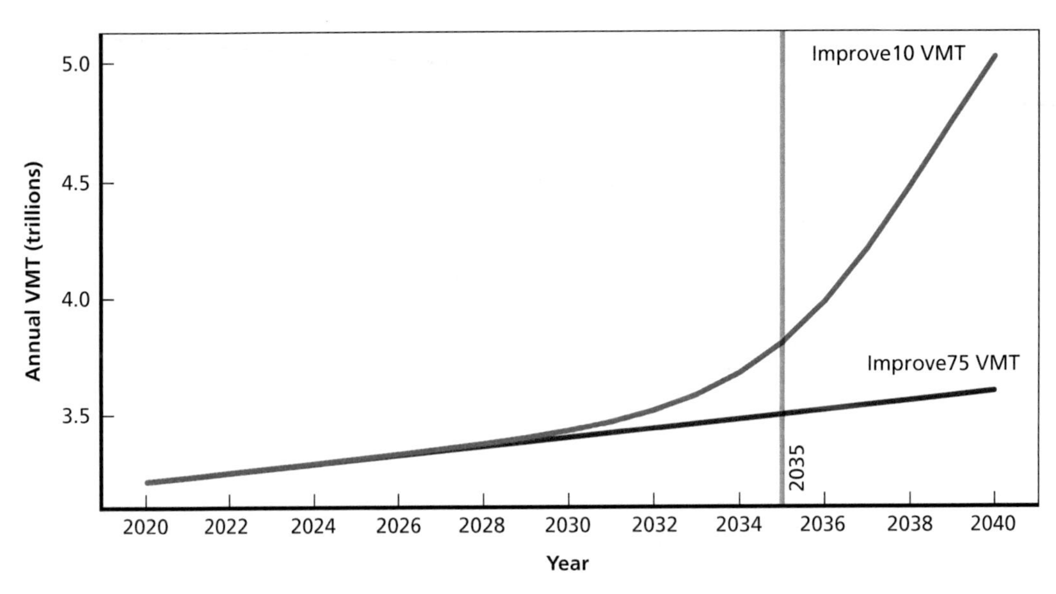

RAND *RR2150-4.4*

Figure 4.5
Improvement in Fatality Rate in Case 117

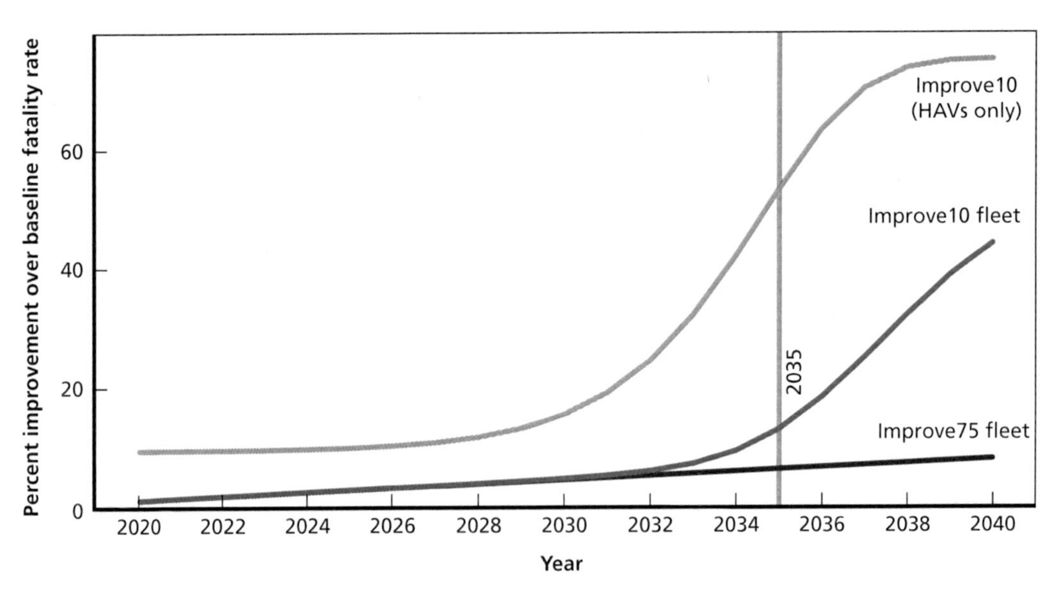

RAND *RR2150-4.5*

still not highly automated, and because there are older HAVs still in operation with performance that does not meet the state of the art. The fatality-rate improvement under Improve75 (in red) is even lower, at 7 percent, because no HAVs have been introduced. (It is improving nonetheless because non-HAV fatality rates decrease in this future.) This is the same rate that would be seen in a future without HAVs.

Figure 4.6 shows that the combination of the high VMT and the not-high-enough improvement in fatality rate can increase annual fatalities compared with having no HAVs at all. In case 117, this increase does not occur under the Improve75 policy because the introduction of HAVs is delayed until the fatality rate is significantly better and can offset the impact of higher VMT. The figure also shows that, under an Improve10 policy, early-year increases in fatalities are offset in later years as HAV safety performance improves and the deployment of these improved HAVs grows. By 2040, there are fewer annual fatalities under Improve10 than under Improve75.

To summarize, in the short term (15 years after introduction), more lives are cumulatively saved under an Improve10 policy than an Improve75 or Improve90 policy under nearly all conditions, and those savings can be significant—tens of thousands to hundreds of thousands of lives. The savings are particularly large when HAVs under Improve10 are adopted quickly. Conversely, the more stringent policies save more lives only when HAVs introduced under Improve10 lead to large VMT increases that are not offset by correspondingly rapid improvements in the HAV fatality rate. However, even under these conditions, the short-term life savings under Improve75 and Improve90

Figure 4.6
Annual Fatalities in Case 117

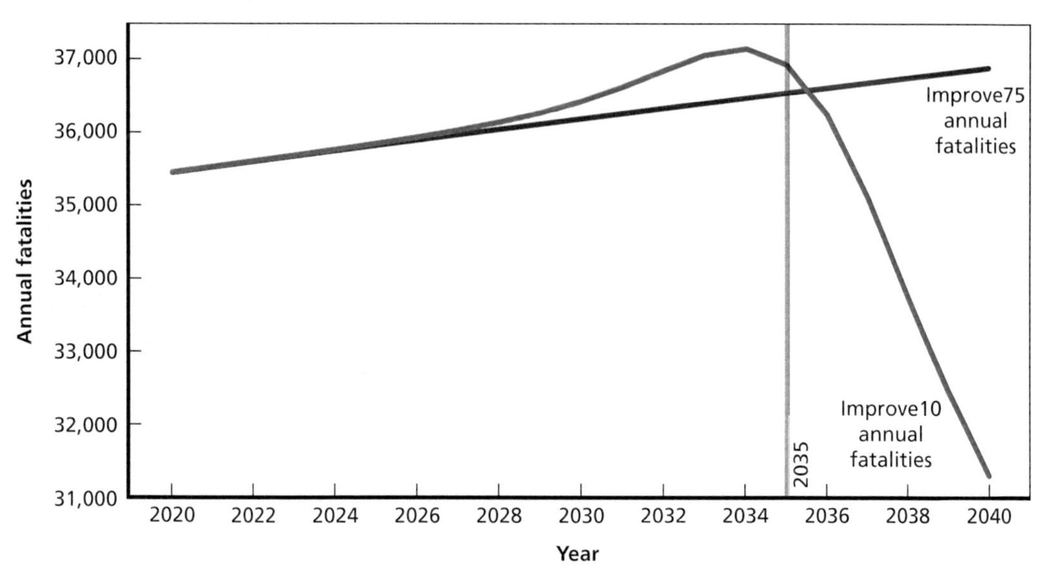

RAND *RR2150-4.6*

policies are relatively small (approximately 3,000 lives over 15 years) and disappear over time as HAV fatality rates continue to improve under an Improve10 policy.

Under What Conditions Are More Lives Saved by Each Policy in the Long Term, and How Large Are Those Savings?

We now focus on the longer-term savings—those occurring 30 years after initial deployment of HAVs under an Improve10 policy. Figure 4.7 shows the difference in cumulative fatalities between Improve10 and Improve75 across the ensemble of cases in the longer term, measured in the year 2050. As in Figure 4.1, positive values (shown in blue) indicate that Improve10 saves more lives cumulatively, while negative values indicate that Improve75 saves more lives cumulatively.

As Figure 4.7 shows, over 30 years, there are more cumulative lives saved under Improve10 than Improve75 under *every* combination of future adoption and performance conditions that we examined. In other words, we could not find a plausible set of conditions in which waiting for significant safety improvements saved more lives in the long run. Moreover, the cost of waiting for those significant improvements can be very large—in some cases, more than half a million lives. This represents

Figure 4.7
Ensemble Difference in Cumulative Lives Saved over 30 Years for 500 Cases, Improve10 Versus Improve75

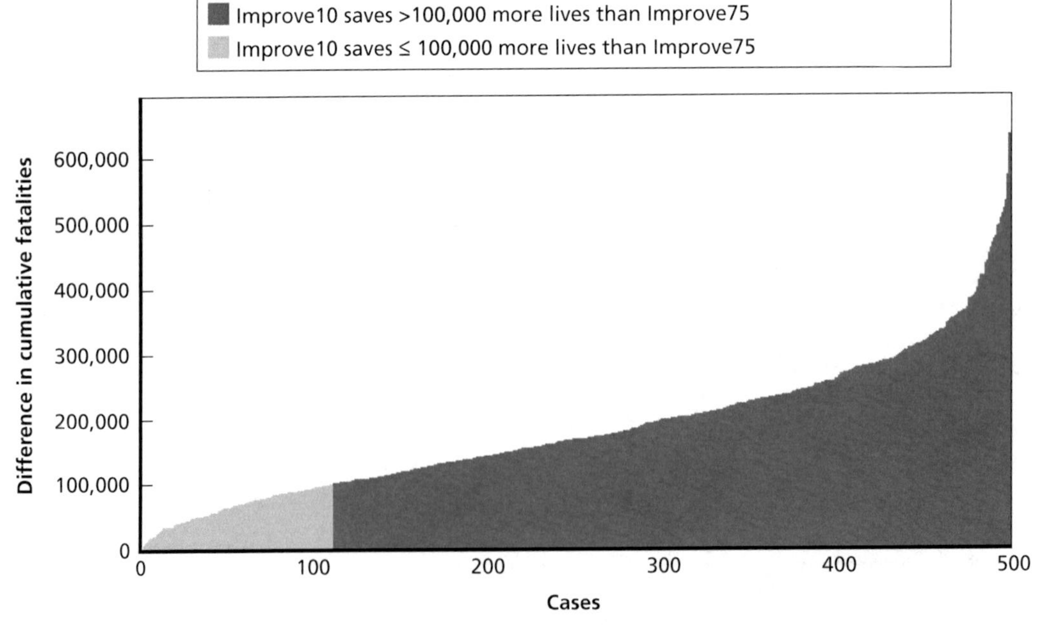

RAND *RR2150-4.7*

up to 83 percent more fatalities than would be seen over the same period under a Improve10 policy.

The difference in life savings between Improve10 and Improve90 follows a very similar pattern (which, for brevity, we do not show in a figure). Improve10 saves more lives in every future examined, and the difference can reach 700,000 lives in the most-extreme cases. Thus, Improve90 has 116 percent more fatalities than would be seen over the same period under an Improve10 policy.

As in the short term, the magnitude of the difference in fatalities is driven primarily by how many more highly automated miles are driven under Improve10 than under Improve75 or Improve90. In the longer term, the difference in VMT and therefore cumulative fatalities is smallest in futures in which HAVs are introduced under both policies nearly simultaneously. Figure 4.8 shows that this generally occurs when (1) the number of miles it takes to achieve full improvement under Improve10 (that is, a 10-percent improvement in the fatality rate over the benchmark rate) is small (horizontal axis) and (2) the same improvement can be achieved equally quickly in laboratory or simulation settings, rather than through deployment, and therefore there is little or no delay in introducing HAVs under an Improve75 or Improve90 policy (vertical axis). As before, futures with a greater difference in fatalities are represented by larger circles in the figure; in this case, the differences repre-

Figure 4.8
Differences in Cumulative Fatalities Between Improve10 and Improve75 or Improve90 over 30 Years Given Two Conditions

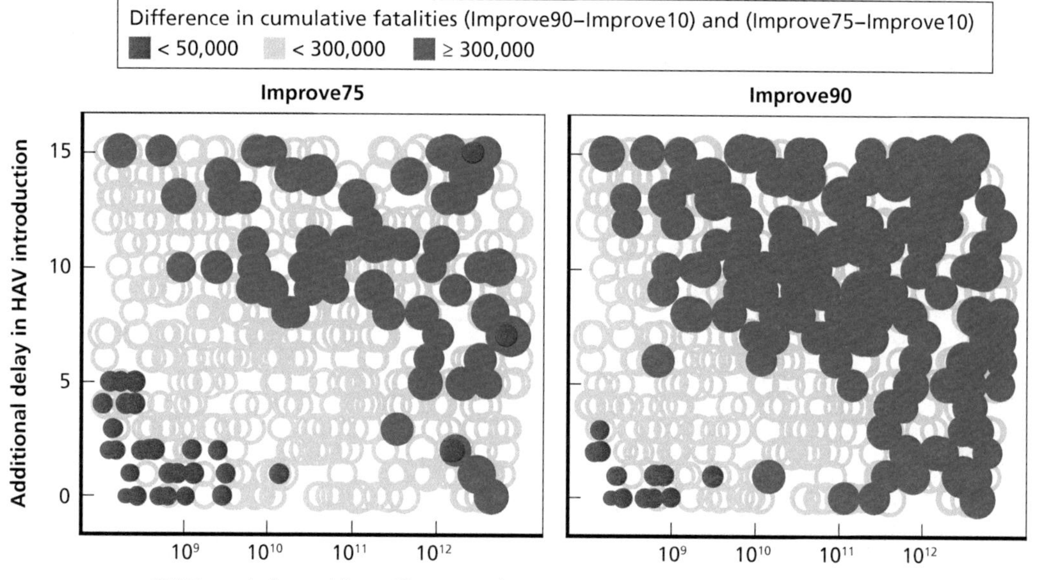

RAND *RR2150-4.8*

sent how many more cumulative fatalities an Improve75 or Improve90 policy leads to over 30 years compared with an Improve10 policy. Futures in which the difference in fatalities is 50,000 or fewer are shown in solid red circles; futures in which the difference is greater than 300,000 are shown in solid blue circles; and futures with differences in between these thresholds are shown in open gray circles. As mentioned earlier, there are more cumulative fatalities under Improve75 and Improve90 than under Improve10 in the long term in all cases.

This reveals that a difference of 50,000 or fewer fatalities occurs when it takes fewer than 10 billion (10^9) miles of driving to improve safety performance and when the additional delay in HAV introduction under Improve75 or Improve90 is five years or less. An increase in fatalities of 300,000 occurs in many cases when it takes 100 billion or more miles of driving and when the additional delay in introduction is ten years or more. Such a large increase can also occur when learning is fast but the delay is large, or vice versa, indicating that there is a trade-off between these factors.

There are, of course, other factors at play that create variation in these outcomes. Factors that allow for more highly automated VMT (such as larger increases in overall VMT and greater use of HAVs) allow for larger differences in life savings from HAVs under Improve10 versus Improve 75. In contrast, poor upgradeability diminishes this difference because legacy vehicles with poorer safety performance remain in operation. Nevertheless, our analysis reveals that these factors are not the primary drivers of different outcomes between these policies.

To summarize, in the longer term, more lives are cumulatively saved under an Improve10 policy than either an Improve75 or Improve90 policy under all combinations of conditions we explored, and those savings can be significant—hundreds of thousands of lives in many cases and more than half a million lives in others. The savings are largest when the introduction of HAVs under Improve75 or Improve90 is significantly delayed because (1) the miles it takes to improve HAVs from better than the average human to nearly perfect is large and (2) the same improvement cannot be achieved equally quickly in laboratory or simulation settings. Savings are smallest when the opposite conditions hold.

CHAPTER FIVE

Policy Implications and Conclusions

Chapter Four presented the analytic results that identify the conditions under which more lives are saved by each policy in the short term and the long term, and how large those savings are. In the short term (within 15 years), more lives are cumulatively saved under a more permissive policy (Improve10) than stricter policies requiring greater safety advancements (Improve75 or Improve90) in nearly all conditions, and those savings can be significant—hundreds of thousands of lives. The savings are largest when HAVs under Improve10 are adopted quickly. Conversely, the more stringent policies save more lives only when the introduction of HAVs would lead to large VMT increases that are not offset by correspondingly rapid reductions in the HAV fatality rate under Improve10. However, even under these conditions, the short-term life savings under the stringent Improve75 or Improve90 policies are relatively small (at most, roughly 3,000 lives cumulatively) and disappear over time as HAV fatality rates continue to improve under an Improve10 policy.

In the long term (within 30 years), more lives are cumulatively saved under an Improve10 policy than either Improve75 or Improve90 policies under *all* combinations of conditions we explored. Those savings can be even larger—in many cases, more than half a million lives. The savings are largest when the introduction of HAVs under Improve75 or Improve90 is significantly delayed because (1) the number of miles it takes to achieve 10-percent improvement in the fatality rate over benchmark under Improve10 is large and (2) the same improvement cannot be achieved equally quickly in laboratory or simulation settings. Savings are smallest when the opposite conditions hold.

This chapter now explores the second and third questions posed by this study:

2. What does the evidence suggest about the conditions that lead to small costs from waiting for technology that is many times safer than human drivers?
3. What does this imply for policies governing the introduction of HAVs for consumer use?

What Does the Evidence Suggest About the Conditions That Lead to Small Costs from Waiting for Technology That Is Many Times Safer Than Human Drivers?

The fact that most futures we examined have high costs of waiting for significantly improved HAVs does not necessarily mean that this is the most likely consequence of waiting, because the futures are not probabilistically generated. This is because the factors that shape fatalities under each policy are deeply uncertain, and probabilities cannot be assigned to any particular outcome. Consistent with RDM, the futures are used to identify the conditions that lead to different policy outcomes, without a priori assumptions about the likelihood of each future.

Given this, it is more appropriate to ask whether there is evidence to suggest that the conditions that lead to a small or no cost of waiting for HAVs that are much better than human drivers are more plausible than those that lead to higher cost. In other words, do we have reason to believe that the pace of HAV improvement will be fast, or that there is little value of postdeployment learning? If so, then one might still favor waiting for HAVs until major safety improvements are achieved.

First, there is no evidence to suggest that improvement of HAVs from just-better-than-average humans to many times better will be quick, whatever the tools and techniques available. Commercial development of the technology began several decades ago and has been under way in earnest for roughly the past decade, with nearly every major automaker developing the technology. While it is unclear at this time whether HAVs today are better or worse than the average human driver, many industry leaders believe that the industry is a long way from reaching significant improvements:

> "[N]one of us in the automobile or [information technology] industries are close to achieving true level 5 autonomy. Collectively, our current prototype autonomous cars can handle many situations. But there are still many others that are beyond current machine competence. It will take many years of machine learning and many more miles than anyone has logged of both simulated and real-world testing to achieve the perfection required for Level 5 autonomy." (Pratt, 2017)

Simultaneously, research on technology improvement across industries suggests that there is a power-law relationship between production and performance: A doubling of production leads to a constant improvement in performance (as measured by cost or other characteristics) (Nagy et al., 2013). A similar relationship may exist for the performance of machine learning algorithms and data sets (Sun et al., 2017). For HAVs, this suggests that achieving gains that some might consider "near perfect" may take much more effort and time than reaching better-than-average human performance, which may itself be still out of reach.

Second, there is no reason to believe that the value of real-world driving to HAV improvement is small. Real-world driving is a huge part of the development effort, as indicated by the 36 companies that have registered to date with the California Department of Motor Vehicles to conduct testing on public roads (California Department of Motor Vehicles, undated). Alternatives—simulations and closed courses—are complementary but cannot replace real-world driving. Thus, there is not a compelling reason to believe that the difference in the timing of introduction of HAVs under the two policies would be small.

In sum, our analysis suggests that a policy of waiting for HAVs to be many times better than human drivers or nearly perfect would be costly in terms of human lives. Just how costly? It may vary significantly—from tens of thousands to hundreds of thousands of lives over time, depending on how the technology and its diffusion evolve.

What Does This Imply for Policies Governing the Introduction of Highly Automated Vehicles for Consumer Use?

In a utilitarian society, our findings would imply that policymakers should allow and developers should deploy HAVs when their safety performance is better than that of the average human driver. However, we do not live in a utilitarian society (Bonnefon, Shariff, and Rahwan, 2016). Thus, for example, a major backlash against a crash caused by even relatively safe HAVs could grind the industry to a halt—resulting in potentially the greatest loss of life over time. As another example, public outcry over technology failures is sometimes the force behind industry making further safety improvements that might otherwise be overlooked.[1]

Instead, our findings suggest that there are real and large costs to waiting for nearly perfect technology and that society—including the public, policymakers, the judicial system, and the transportation industry—must balance the social response to HAV crashes with the rate of HAV crashes under different policy options. The evidence in this report can help stakeholders find a middle ground of HAV performance requirements that may prove to save the most lives overall.

---

[1] Airbags offer such an example. Airbags helped save the lives of many adult male passengers but also injured and killed some smaller-statured passengers. The public and policymakers demanded improvements to the technology, spurring the development of the smarter airbags found in vehicles today (Houston and Richardson, 2000).

References

Abraham, Hillary, Bryan Reimer, Bobbie Seppelt, Craig Fitzgerald, Bruce Mehle, and Joseph F. Coughlin, *Consumer Interest in Automation: Preliminary Observations Exploring a Year's Change*, Cambridge, Mass.: Massachusetts Institute of Technology AgeLab, White Paper 2017-2, 2017. As of September 12, 2017:
http://agelab.mit.edu/sites/default/files/MIT%20-%20NEMPA%20White%20Paper%20FINAL.pdf

Anderson, James M., Nidhi Kalra, Karlyn D. Stanley, Paul Sorensen, Constantine Samaras, and Oluwatobi A. Oluwatola, *Autonomous Vehicle Technology: A Guide for Policymakers*, Santa Monica, Calif.: RAND Corporation, RR-443-2-RC, 2016. As of January 24, 2016:
http://www.rand.org/pubs/research_reports/RR443-2.html

Bansal, Prateek, and Kara M. Kockelman, "Forecasting Americans' Long-Term Adoption of Connected and Autonomous Vehicle Technologies," *Transportation Research Part A: Policy and Practice*, Vol. 95, January 2017, pp. 49–63.

Blincoe, Lawrence, Ted R. Miller, Eduard Zaloshnja, and Bruce A. Lawrence, *The Economic and Societal Impact of Motor Vehicle Crashes 2010 (Revised)*, Washington, D.C.: U.S. Department of Transportation, DOT HS 812 013, May 2015. As of March 3, 2016:
http://www-nrd.nhtsa.dot.gov/pubs/812013.pdf

Bonnefon, Jean-François, Azim Shariff, and Iyad Rahwan, "The Social Dilemma of Autonomous Vehicles," *Science*, Vol. 352, No. 6293, 2016, pp. 1573–1576.

Boston Consulting Group, "By 2030, 25% of Miles Driven in US Could Be in Shared Self-Driving Electric Cars," press release, April 10, 2017. As of September 12, 2017:
https://www.bcg.com/d/press/10april2017-future-autonomous-electric-vehicles-151076

Bureau of Transportation Statistics, "Table 2-17: Motor Vehicle Safety Data," in *National Transportation Statistics*, Washington, D.C.: U.S. Department of Transportation, April 2016. As of January 31, 2017:
https://www.rita.dot.gov/bts/sites/rita.dot.gov.bts/files/publications/national_transportation_statistics/html/table_02_17.html

California Department of Motor Vehicles, "Testing of Autonomous Vehicles," web page, undated. As of September 12, 2017:
https://www.dmv.ca.gov/portal/dmv/detail/vr/autonomous+/testing

Casualty Actuarial Society, *Restating the National Highway Transportation Safety Administration's National Motor Vehicle Crash Causation Survey for Automated Vehicles*, Casualty Actuarial Society E-Forum, Vol. 1, Fall 2014.

Collins, Susan M., "Chairman Susan M. Collins Opening Statement," hearing on "The Automated and Self-Driving Vehicle Revolution: What Is the Role of Government?" U.S. Senate, Committee on Appropriations, Subcommittee on Transportation, Housing and Urban Development, and Related Agencies, November 16, 2016. As of September 8, 2017:
https://www.appropriations.senate.gov/imo/media/doc/111616-Chairman-Collins-Opening-Statement-21.pdf

DeCicco, John M., *A Fuel Efficiency Horizon for U.S. Automobiles*, September 2010. As of September 12, 2017:
https://deepblue.lib.umich.edu/bitstream/handle/2027.42/78178/DeCicco_AutoEfficiencyHorizon_Sept2010.pdf?sequence=1

Dietvorst, Berkeley J., Joseph P. Simmons, and Cade Massey, "Algorithm Aversion: People Erroneously Avoid Algorithms After Seeing Them Err," *Journal of Experimental Psychology: General*, July 6, 2014.

———, "Overcoming Algorithm Aversion: People Will Use Imperfect Algorithms If They Can (Even Slightly) Modify Them," *Management Science*, November 4, 2016.

Dingus, T. A., F. Guo, S. Lee, J. F. Antin, M. Perez, M. Buchanan-King, and J. Hankey, "Driver Crash Risk Factors and Prevalence Evaluation Using Naturalistic Driving Data," *Proceedings of the National Academy of Sciences*, Vol. 113, No. 10, 2016, pp. 2636–2641.

Dixon, Lloyd, Robert J. Lempert, Tom LaTourrette, and Robert T. Reville, *The Federal Role in Terrorism Insurance: Evaluating Alternatives in an Uncertain World*, Santa Monica, Calif.: RAND Corporation, MG-679-CTRMP, 2007. As of September 12, 2017:
http://www.rand.org/pubs/monographs/MG679.html

Eccles, K., F. Gross, M. Liu, and F. Council, *Crash Data Analyses for Vehicle-to-Infrastructure Communications for Safety Applications*, Washington, D.C.: Federal Highway Administration, DTFH61-06-C-00013, 2012.

Energy Information Administration, *Annual Energy Outlook 2006*, Washington, D.C.: U.S. Department of Energy, DOE/EIA-3083, February 2006. As of September 12, 2017:
http://maecourses.ucsd.edu/MAE119/WI_2015/PDF-PublishedDocuments/US_EIAAnnualEnergyOutlook-2006.pdf

———, "Appendix A: Reference Case," in *Annual Energy Outlook 2017*, Washington, D.C.: U.S. Department of Energy, January 2017a. As of September 12, 2017:
https://www.eia.gov/outlooks/aeo/pdf/appa.pdf

———, *Study of the Potential Energy Consumption Impacts of Connected and Automated Vehicles*, Washington, D.C.: U.S. Department of Energy, March 2017b. As of September 12, 2017:
https://www.eia.gov/analysis/studies/transportation/automated/pdf/automated_vehicles.pdf

Fagnant, Daniel J., and Kara Kockelman, "Preparing a Nation for Autonomous Vehicles: Opportunities, Barriers and Policy Recommendations," *Transportation Research Part A: Policy and Practice*, Vol. 77, July 2015, pp. 167–181.

Federal Highway Administration, *FHWA Forecasts of Vehicle Miles Traveled (VMT): Spring 2016*, Washington, D.C.: U.S. Department of Transportation, May 2, 2016.

———, *FHWA Forecasts of Vehicle Miles Traveled (VMT): Spring 2017*, Washington, D.C.: U.S. Department of Transportation, May 4, 2017. As of January 17, 2017:
https://www.fhwa.dot.gov/policyinformation/tables/vmt/vmt_forecast_sum.pdf

FHWA—*See* Federal Highway Administration.

Fischbach, Jordan R., Kyle Siler-Evans, Devin Tierney, Michael T. Wilson, Lauren M. Cook, and Linnea Warren May, *Robust Stormwater Management in the Pittsburgh Region: A Pilot Study*, Santa Monica, Calif.: RAND Corporation, RR-1673-MCF, 2017. As of September 12, 2017: https://www.rand.org/pubs/research_reports/RR1673.html

Fischhoff, Baruch, Paul Slovic, Sarah Lichtenstein, Stephen Read, and Barbara Combs, "How Safe Is Safe Enough? A Psychometric Study of Attitudes Towards Technological Risks and Benefits," *Policy Sciences*, Vol. 9, No. 2, 1978, pp. 127–152.

Ford Motor Company, "Ford Targets Fully Autonomous Vehicle for Ride Sharing in 2021; Invests in New Tech Companies, Doubles Silicon Valley Team," press release, August 16, 2016. As of September 12, 2017:
https://media.ford.com/content/fordmedia/fna/us/en/news/2016/08/16/ford-targets-fully-autonomous-vehicle-for-ride-sharing-in-2021.html

Fraade-Blanar, Laura, and Nidhi Kalra, *Autonomous Vehicles and Federal Safety Standards: An Exemption to the Rule?* Santa Monica, Calif.: RAND Corporation, PE-258-RC, 2017. As of September 12, 2017:
https://www.rand.org/pubs/perspectives/PE258.html

Funke, James, Gowrishankar Srinivasan, Raja Ranganathan, and August Burgett, *Safety Impact Methodology (SIM): Application and Results of the Advanced Crash Avoidance Technologies (ACAT) Program*, Paper No. 11-0367, 2011.

Gomes, Lee, "Hidden Obstacles for Google's Self-Driving Cars: Impressive Progress Hides Major Limitations of Google's Quest for Automated Driving," *MIT Technology Review*, August 28, 2014. As of March 3, 2016:
https://www.technologyreview.com/s/530276/hidden-obstacles-for- googles-self-driving-cars/

Google Auto LLC, *Google Self-Driving Car Project: Monthly Report*, Menlo Park, Calif., January 2016. As of September 12, 2017:
https://static.googleusercontent.com/media/www.google.com/en//selfdrivingcar/files/reports/report-0116.pdf

Gordon, T., H. Sardar, D. Blower, M. Ljung Aust, Z. Bareket, M. Barnes, A. Blankespoor, I. Isaksson-Hellman, J. Ivarsson, B. Juhas, K. Nobukawa, and H. Theander, *Advanced Crash Avoidance Technologies (ACAT) Program: Final Report of the Volvo-Ford-UMTRI Project: Safety Impact Methodology for Lane Departure Warning—Method Development and Estimation of Benefits*, Washington, D.C.: U.S. Department of Transportation, DOT HS 811 405, October 2010.

Groves, David G., Martha Davis, Robert Wilkinson, and Robert Lempert, "Planning for Climate Change in the Inland Empire: Southern California," *Water Resources IMPACT*, Vol. 10, No. 4, July 2008, pp. 14–17. As of September 12, 2017:
http://www.jstor.org/stable/wateresoimpa.10.4.0014

Groves, David G., Jordan R. Fischbach, Evan Bloom, Debra Knopman, and Ryan Keefe, *Adapting to a Changing Colorado River: Making Future Water Deliveries More Reliable Through Robust Management Strategies*, Santa Monica, Calif.: RAND Corporation, RR-242-BOR, 2013. As of October 20, 2017:
http://www.rand.org/pubs/research_reports/RR242.html

Groves, David G., and Robert J. Lempert, "A New Analytic Method for Finding Policy-Relevant Scenarios," *Global Environmental Change*, Vol. 17, No. 1, 2007, pp. 73–85.

Harper, C., C. Hendrickson, S. Mangones, and S. Constantine, "Estimating Potential Increases in Travel with Autonomous Vehicles for the Non-Driving, Elderly and People with Travel-Restrictive Medical Conditions," *Transportation Research Part C: Emerging Technologies*, Vol. 72, November 2016.

Harper, Corey D., Chris T. Hendrickson, and Constantine Samaras, "Cost and Benefit Estimates of Partially-Automated Vehicle Collision Avoidance Technologies," *Accident Analysis and Prevention*, Vol. 95, Part A, October 2016, pp. 104–115.

Highway Loss Data Institute, "Predicted Availability of Safety Features on Registered Vehicles," bulletin, Arlington, Va., Vol. 28, No. 26, April 2012. As of September 12, 2017:
http://www.iihs.org/media/db4aeba1-6209-4382-9ef2-275443fcccea/_LdQUQ/HLDI%20Research/Bulletins/hldi_bulletin_28.26.pdf

Houston, David J., and Lilliard E. Richardson, Jr., "The Politics of Air Bag Safety: A Competition Among Problem Definitions," *Policy Studies Journal*, Vol. 28, No. 3, August 2000, pp. 485–501.

Hsu, Jeremy, "When It Comes to Safety, Autonomous Cars Are Still 'Teen Drivers,'" *Scientific American*, January 18, 2017. As of September 12, 2017:
https://www.scientificamerican.com/article/when-it-comes-to-safety-autonomous-cars-are-still-teen-drivers1/

Jermakian, Jessica S., "Crash Avoidance Potential of Four Passenger Vehicle Technologies," *Accident Analysis and Prevention*, Vol. 43, No. 3, May 2011, pp. 732–740.

Jutila, Sakari T., "Dynamic Modeling of Adoption, Rejection and Life Cycles of Innovations," in Qifan Wang and Robern Eberlein, eds., *The 5th International Conference of the System Dynamics Society*, Shanghai, 1987. As of September 12, 2017:
http://www.systemdynamics.org/conferences/1987/proceed/jutil274.pdf

Kalra, Nidhi, *Challenges and Approaches to Realizing Autonomous Vehicle Safety*, testimony submitted to the House Energy and Commerce Committee, Subcommittee on Digital Commerce and Consumer Protection, Santa Monica, Calif.: RAND Corporation, CT-463, February 14, 2017. As of September 12, 2017:
https://www.rand.org/pubs/testimonies/CT463.html

Kalra, Nidhi, and David G. Groves, *RAND Model of Autonomous Vehicle Safety (MAVS): Model Documentation*, Santa Monica, Calif.: RAND Corporation, RR-1902-RC, 2017. As of October 1, 2017:
https://www.rand.org/pubs/research_reports/RR1902.html

Kalra, Nidhi, S. Hallegatte, Robert Lempert, Casey Brown, Adrian Fozzard, Stuart Gill, and Ankur Shah, *Agreeing on Robust Decisions: New Processes for Decision Making Under Deep Uncertainty*, World Bank Policy Research Working Paper No. 6906, June 1, 2014. As of September 12, 2017:
http://ssrn.com/abstract=2446310

Kalra, Nidhi, and Susan Paddock, *Driving to Safety: How Many Miles of Driving Would It Take to Demonstrate Autonomous Vehicle Reliability?* Santa Monica, Calif.: RAND Corporation, RR-1478-RC, 2016. As of September 12, 2017:
http://www.rand.org/pubs/research_reports/RR1478.html

Koopman, Philip, and Michael Wagner, "Autonomous Vehicle Safety: An Interdisciplinary Challenge," *IEEE Intelligent Transportation Systems Magazine*, Vol. 9, No. 1, Spring 2017. As of September 12, 2017:
http://ieeexplore.ieee.org/document/7823109/

Kutila, Matti, Pasi Pyykönen, Werner Ritter, Oliver Sawade, and Bernd Schäufele, "Automotive LIDAR Sensor Development Scenarios for Harsh Weather Conditions," *2016 IEEE 19th International Conference on Intelligent Transportation Systems*, November 2016, pp. 265–270.

Lemmer, Karsten, "Effectively Ensuring Automated Driving," presentation at the VDA Technical Congress, Berlin, April 6, 2017. As of September 12, 2017:
http://www.pegasus-projekt.info/files/tmpl/pdf/PEGASUS_VDA_techn.congress_EN.pdf

Lempert, Robert, and Nidhi Kalra, *Managing Climate Risks in Developing Countries with Robust Decision Making*, Washington D.C.: World Resources Institute, 2011.

Lempert, Robert J., Steven W. Popper, and Steven C. Bankes, *Shaping the Next One Hundred Years: New Methods for Quantitative, Long-Term Policy Analysis*, Santa Monica, Calif.: RAND Corporation, MR-1626-RPC, 2003. As of September 12, 2017:
https://www.rand.org/pubs/monograph_reports/MR1626.html

Lempert, Robert J., Steven W. Popper, David G. Groves, Nidhi Kalra, Jordan R. Fischbach, Steven C. Bankes, Benjamin P. Bryant, Myles T. Collins, Klaus Keller, Andrew Hackbarth, Lloyd Dixon, Tom LaTourrette, Robert T. Reville, Jim W. Hall, Christophe Mijere, and David J. McInerney, "Making Good Decisions Without Predictions: Robust Decision Making for Planning Under Deep Uncertainty," Santa Monica, Calif.: RAND Corporation, RB-9701, 2013. As of September 12, 2017:
http://www.rand.org/pubs/research_briefs/RB9701/index1.html

Li, Tianxin, and Kara M. Kockelman, *Valuing the Safety Benefits of Connected and Automated Vehicle Technologies*, Transportation Research Board 95th Annual Meeting, 2016.

Litman, Todd, *Autonomous Vehicle Implementation Predictions: Implications for Transport Planning*, Victoria, British Columbia, Canada: Victoria Transport Policy Institute, January 2, 2017.

Milakis, Dimitris, Bart van Arem, and Bert van Wee, "Policy and Society Related Implications of Automated Driving: A Review of Literature and Directions for Future Research," *Journal of Intelligent Transportation Systems*, Vol. 21, No. 4, 2017, pp. 324–348.

Musk, Elon, "Tesla Press Conference for the Autopilot v7.0 Software," October 14, 2015. As of December 15, 2016:
https://www.youtube.com/watch?v=73_Qjez1MbI

Nagy, Béla, J. Doyne Farmer, Quan M. Bui, and Jessika E. Trancik, "Statistical Basis for Predicting Technological Progress," *PLoS ONE*, Vol. 8, No. 2, e52669, 2013.

Najm, Wassim G., Samuel Toma, and John Brewer, *Depiction of Priority Light-Vehicle Pre-Crash Scenarios for Safety Applications Based on Vehicle-to-Vehicle Communications*, Washington, D.C.: U.S. Department of Transportation, DOT HS 811 732, April 2013. As of September 12, 2017:
https://ntl.bts.gov/lib/47000/47400/47497/DOT-VNTSC-NHTSA-11-12.pdf

National Highway Traffic Safety Administration, *National Motor Vehicle Crash Causation Survey*, Washington, D.C.: U.S. Department of Transportation, DOT HS 811 059, 2008.

———, *Drowsy Driving*, Washington, D.C.: U.S. Department of Transportation, DOT HS 811 449, March 2011. As of July 28, 2017:
https://crashstats.nhtsa.dot.gov/Api/Public/ViewPublication/811449

———, *Critical Reasons for Crashes Investigated in the National Motor Vehicle Crash Causation Survey*, Washington, D.C.: U.S. Department of Transportation, DOT HS 812 115, February 2015. As of September 12, 2017:
https://crashstats.nhtsa.dot.gov/Api/Public/ViewPublication/812115

———, *2015 Motor Vehicle Crashes: Overview*, Washington, D.C.: U.S. Department of Transportation, DOT HS 812 318, August 2016a. As of January 18, 2017:
https://crashstats.nhtsa.dot.gov/Api/Public/ViewPublication/812318

———, *Federal Automated Vehicle Policy: Accelerating the Next Revolution in Roadway Safety*, Washington, D.C.: U.S. Department of Transportation, September 2016b. As of September 12, 2017:
https://www.transportation.gov/sites/dot.gov/files/docs/AV%20policy%20guidance%20PDF.pdf

———, *Alcohol-Impaired Driving*, Washington, D.C.: U.S. Department of Transportation, DOT HS 812 350, December 2016c. As of July 28, 2017:
https://crashstats.nhtsa.dot.gov/Api/Public/Publication/812350

———, *Distracted Driving 2014*, Washington, D.C.: U.S. Department of Transportation, DOT HS 812 260, December 2016d. As of July 28, 2017:
https://crashstats.nhtsa.dot.gov/Api/Public/ViewPublication/812260

———, *2016 Fatal Motor Vehicle Crashes: Overview*, Washington, D.C.: U.S. Department of Transportation, DOT HS 812 456, October 2017. As of October 25, 2017:
https://crashstats.nhtsa.dot.gov/Api/Public/Publication/812456

NHTSA—*See* National Highway Transportation Safety Administration.

Nowakowski, Christopher, Steven E. Shladover, and Ching-Yao Chan, "Determining the Readiness of Automated Driving Systems for Public Operation: Development of Behavioral Competency Requirements," *Transportation Research Record: Journal of the Transportation Research Board*, Vol. 2559, 2016, pp. 65–72.

Otway, Harry J., and Detlof von Winterfeldt, "Beyond Acceptable Risk: On the Social Acceptability of Technologies," *Policy Sciences*, Vol. 14, No. 3, 1982, pp. 247–256.

Perez, M., L. S. Angell, J. Hankey, R. K. Deering, R. E. Llaneras, C. A. Green, M. L. Neurauter, and J. F. Antin, *Advanced Crash Avoidance Technologies (ACAT) Program—Final Report of the GM–VTTI Backing Crash Countermeasures Project*, Washington, D.C.: U.S. Department of Transportation, DOT HS 811 452, August 2011.

Petit, J., and S. E. Shladover, "Potential Cyberattacks on Automated Vehicles," *IEEE Transactions on Intelligent Transportation Systems*, Vol. 16, No. 2, 2015, pp. 546–556.

Popper, Steven W., Claude Berrebi, James Griffin, Thomas Light, Endy M. Daehner, and Keith Crane, *Natural Gas and Israel's Energy Future: Near-Term Decisions from a Strategic Perspective*, Santa Monica, Calif.: RAND Corporation, MG-927-YSNFF, 2009. As of September 12, 2017:
http://www.rand.org/pubs/monographs/MG927

Pratt, Gill, "2017 Consumer Electronics Show (CES2017) Press Conference," Toyota, January 4, 2017.
As of September 12, 2017:
http://corporatenews.pressroom.toyota.com/article_display.cfm?article_id=5878

Rau, Paul, Mikio Yanagisawa, and Wassim G. Najm, *Target Crash Population of Automated Vehicles*, Paper No. 15-0430, 2015.

Roose, Kevin, "As Self-Driving Cars Near, Washington Plays Catch-Up," *New York Times*, July 21, 2017. As of September 12, 2017:
https://www.nytimes.com/2017/07/21/technology/self-driving-cars-washington-congress.html

Ross, Philip E., "CES 2017: Nvidia and Audi Say They'll Field a Level 4 Autonomous Car in Three Years," *IEEE Spectrum*, January 5, 2017. As of September 12, 2017:
http://spectrum.ieee.org/cars-that-think/transportation/self-driving/nvidia-ceo-announces

SAE International, *Surface Vehicle Recommended Practice: Taxonomy and Definitions for Terms Related to Driving Automation Systems for On-Road Motor Vehicles*, Warrendale, Pa., J2016, September 2016.

Saltelli, A., K. Chan, and E. M. Scott, *Sensitivity Analysis*, New York: Wiley and Sons, 2000.

Shladover, Steven, "The Truth About 'Self-Driving' Cars," *Scientific American*, December 2016. As of September 12, 2017:
https://www.scientificamerican.com/article/the-truth-about-ldquo-self-driving-rdquo-cars/

Sjöberg, Lennart, "Factors in Risk Perception," *Risk Analysis*, Vol. 20, No. 1, 2000, pp. 1–12.

Slovic, P., "Perception of Risk," *Science*, Vol. 236, 1987, pp. 280–285.

———, ed., *The Perception of Risk*, New York: Earthscan, 2000.

Slovic, Paul, and Ellen Peters, "Risk Perception and Affect," *Current Directions in Psychological Science*, Vol. 15, No. 6, 2006, pp. 322–325.

Starr, Chauncey, "Social Benefit Versus Technological Risk," *Science*, Vol. 165, No. 3899, September 1969, pp. 1232–1238.

Sun, Chen, Abhinav Shrivastava, Saurabh Singh, and Abhinav Gupta, "Revisiting Unreasonable Effectiveness of Data in Deep Learning Era," arXiv, July 2017. As of September 12, 2017: https://arxiv.org/abs/1707.02968

Tesla, *Tesla Third Quarter 2016 Update*, Palo Alto, Calif., 2016. As of September 12, 2017: http://files.shareholder.com/downloads/ABEA-4CW8X0/2238129554x0x913801/F9E5C36A-AFDD-4FF2-A375-ED9B0F912622/Q3_16_Update_Letter_-_final.pdf

Walden, Greg, "Opening Statement of Chairman Greg Walden," hearing on "Self-Driving Cars: Road to Deployment," U.S. House of Representatives, Energy and Commerce Committee, Subcommittee on Digital Commerce and Consumer Protection, February 14, 2017. As of September 8, 2017:
http://docs.house.gov/meetings/IF/IF17/20170214/105548/HHRG-115-IF17-MState-W000791-20170214.pdf

Waymo, "On the Road," web page, undated. As of September 12, 2017: https://waymo.com/ontheroad/

Winkle, Thomas, "Safety Benefits of Automated Vehicles: Extended Findings from Accident Research for Development, Validation and Testing in Autonomous Driving," in Markus Maurer, J. Christian Gerdes, Barbara Lenz, and Hermann Winner, eds., *Autonomous Driving: Technical, Legal and Social Aspects*, London: SpringerOpen, 2015, pp 335–364. As of September 12, 2017: https://link.springer.com/chapter/10.1007/978-3-662-48847-8_17/fulltext.html

U.S. Department of Energy, *The Transforming Mobility Ecosystem: Enabling an Energy-Efficient Future*, Washington, D.C., DOE/EE-1489, January 2017. As of September 12, 2017: https://energy.gov/sites/prod/files/2017/01/f34/The%20Transforming%20Mobility%20Ecosystem-Enabling%20an%20Energy%20Efficient%20Future_0117_1.pdf

U.S. Department of Transportation, "U.S. Department of Transportation Designates 10 Automated Vehicle Proving Grounds to Encourage Testing of New Technologies," press release, January 19, 2017. As of September 12, 2017:
https://www.transportation.gov/briefing-room/dot1717

U.S. Senate, Self-Drive Act, H.R. 3388, September 7, 2017. As of September 12, 2017: https://www.congress.gov/bill/115th-congress/house-bill/3388/text

Zhao, Ding, and Huei Peng, *From the Lab to the Street: Solving the Challenge of Accelerating Automated Vehicle Testing*, Ann Arbor, Mich.: University of Michigan, Mcity, May 2017. As of September 12, 2017:
https://mcity.umich.edu/wp-content/uploads/2017/05/Mcity-White-Paper_Accelerated-AV-Testing.pdf